Wissenschaftliche Reihe Fahrzeugtechnik Universität Stuttgart

Reihe herausgegeben von
Michael Bargende, Stuttgart, Deutschland
Hans-Christian Reuss, Stuttgart, Deutschland
Jochen Wiedemann, Stuttgart, Deutschland

Das Institut für Verbrennungsmotoren und Kraftfahrwesen (IVK) an der Universität Stuttgart erforscht, entwickelt, appliziert und erprobt, in enger Zusammenarbeit mit der Industrie, Elemente bzw. Technologien aus dem Bereich moderner Fahrzeugkonzepte. Das Institut gliedert sich in die drei Bereiche Kraftfahrwesen, Fahrzeugantriebe und Kraftfahrzeug-Mechatronik. Aufgabe dieser Bereiche ist die Ausarbeitung des Themengebietes im Prüfstandsbetrieb, in Theorie und Simulation. Schwerpunkte des Kraftfahrwesens sind hierbei die Aerodynamik, Akustik (NVH), Fahrdynamik und Fahrermodellierung, Leichtbau, Sicherheit, Kraftübertragung sowie Energie und Thermomanagement – auch in Verbindung mit hybriden und batterieelektrischen Fahrzeugkonzepten. Der Bereich Fahrzeugantriebe widmet sich den Themen Brennverfahrensentwicklung einschließlich Regelungs- und Steuerungskonzeptionen bei zugleich minimierten Emissionen, komplexe Abgasnachbehandlung, Aufladesysteme und -strategien, Hybridsysteme und Betriebsstrategien sowie mechanisch-akustischen Fragestellungen. Themen der Kraftfahrzeug-Mechatronik sind die Antriebsstrangregelung/Hybride, Elektromobilität, Bordnetz und Energiemanagement, Funktions- und Softwareentwicklung sowie Test und Diagnose. Die Erfüllung dieser Aufgaben wird prüfstandsseitig neben vielem anderen unterstützt durch 19 Motorenprüfstände, zwei Rollenprüfstände, einen 1:1-Fahrsimulator, einen Antriebsstrangprüfstand, einen Thermowindkanal sowie einen 1:1-Aeroakustikwindkanal. Die wissenschaftliche Reihe „Fahrzeugtechnik Universität Stuttgart" präsentiert über die am Institut entstandenen Promotionen die hervorragenden Arbeitsergebnisse der Forschungstätigkeiten am IVK.

Reihe herausgegeben von

Prof. Dr.-Ing. Michael Bargende
Lehrstuhl Fahrzeugantriebe
Institut für Verbrennungsmotoren und
Kraftfahrwesen, Universität Stuttgart
Stuttgart, Deutschland

Prof. Dr.-Ing. Hans-Christian Reuss
Lehrstuhl Kraftfahrzeugmechatronik
Institut für Verbrennungsmotoren und
Kraftfahrwesen, Universität Stuttgart
Stuttgart, Deutschland

Prof. Dr.-Ing. Jochen Wiedemann
Lehrstuhl Kraftfahrwesen
Institut für Verbrennungsmotoren und
Kraftfahrwesen, Universität Stuttgart
Stuttgart, Deutschland

Weitere Bände in der Reihe http://www.springer.com/series/13535

Andreas Kächele

Turbocharger Integration into Multidimensional Engine Simulations to Enable Transient Load Cases

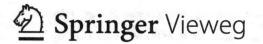
Springer Vieweg

Andreas Kächele
Institute of Internal Combustion Engines
and Automotive Engineering (IVK)
University of Stuttgart
Stuttgart, Germany

Dissertation, University of Stuttgart, 2019

D93

ISSN 2567-0042　　　　　　　ISSN 2567-0352　(electronic)
Wissenschaftliche Reihe Fahrzeugtechnik Universität Stuttgart
ISBN 978-3-658-28785-6　　　　ISBN 978-3-658-28786-3　(eBook)
https://doi.org/10.1007/978-3-658-28786-3

This Springer Vieweg imprint is published by the registered company Springer Fachmedien Wiesbaden GmbH part of Springer Nature.
The registered company address is: Abraham-Lincoln-Str. 46, 65189 Wiesbaden, Germany

Preface

This work was realized during my tenure as a research associate at the Institute of Internal Combustion Engines and Automotive Engineering (IVK) at the University of Stuttgart under the supervision of Prof. Dr.-Ing. M. Bargende. Within this time I have been given the chance to work on numerous interesting projects, while continuously pursuing the aim of enabling 3D-CFD simulations for turbocharged engines.

My deep gratitude goes to Prof. Dr.-Ing. M. Bargende for the opportunity and the continuous extensive scientific support. I also want to thank Prof. Dr.-Ing. P. Eilts for being the co-referent of this work and sharing his professional opinion starting from the very first publication.

I am extremely grateful to all my colleagues, especially Dr. Marco Chiodi, arousing my interest in virtual engine development and the inherent problem solving and interdisciplinary thinking. Mr. Oliver Mack (aka Meshelangelo) has shown great patience and support when introducing me into the generation of structured cylinder meshes. Thanks also to Dr. Marlene Wentsch, Francesco Cupo, Antonino Vacca and all other colleagues at the IVK/FKFS for the friendship and pleasant collaboration during many projects.

I would like to thank Dr. Donatus Wichelhaus for his confidence, scientific opinion and the interesting thermodynamic discussions throughout many joint projects, as well as Dr. Daniel Koch from the TU München. Advice and insight given by Mr. Alexander Fürschuss, regarding the engine design process, have always been of great value for me.

My appreciation also extends to Tom Heuer from BorgWarner for providing detailed information about the turbocharger used in this work and to Dr. Markus Schatz from the ITSM, Stuttgart for providing hot gas test bench data.

Further, I wish to acknowledge the support from the Friedrich und Elisabeth Boysen Foundation and their engagement for environmental technology.

Stuttgart   Andreas Kächele

Contents

Preface V
Figures IX
Tables XI
Abbreviations XIII
Symbols XV
Abstract XIX
Kurzfassung XXI

1   Introduction 1

2   Fundamentals 5
    2.1   The Real Working Process Analysis 6
    2.2   1D-Simulation 7
    2.3   3D-CFD Simulation 8
          2.3.1   Fundamental Equations 9
          2.3.2   Mass Conservation 9
          2.3.3   Momentum Conservation 10
          2.3.4   Energy Conservation 10
          2.3.5   Turbulence 11
    2.4   Turbocharger 13
          2.4.1   Turbine and Compressor Characterization 13
          2.4.2   Turbocharged Internal Combustion Engines 20
          2.4.3   Turbocharger under Transient Conditions 22
    2.5   Virtual Engine Development 26

3   Simulation Environments 29
    3.1   QuickSim 29
          3.1.1   Methodology and Approach 29
          3.1.2   Engine Specific Models 32
    3.2   STAR-CCM+ 35
    3.3   GT-Power 36

4   Turbocharger Integration in QuickSim **37**

4.1   State of the Art for Turbocharged Engines 37

4.2   Benefits of an Integrated Turbocharger 38

4.3   Requirements for an Integrated Turbocharger Model................ 41

4.4   Approaches for an Integrated Turbocharger 43

5   The Chosen Approach: 0D-Turbocharger **47**

5.1   General Modeling and Implementation 47

5.2   Additional Models 51

    5.2.1   Compressor Surge 51

    5.2.2   Reversed Mass Flow 53

    5.2.3   Turbocharger Shaft Friction Model........................... 56

6   Validation by means of a Virtual Hot Gas Test Bench **59**

6.1   Test Bench Setup and Simulation Environments 59

6.2   Artificial Pressure Traces 63

6.3   Stationary Flow on the Virtual Hot Gas Test Bench 64

6.4   Instationary Flow on the Virtual Hot Gas Test Bench 66

    6.4.1   Influence of Pulse Frequency 67

    6.4.2   Influence of Pulse Amplitude 73

    6.4.3   Influence of Turbocharger Shaft Speed....................... 76

    6.4.4   Simulation Stability 78

    6.4.5   Comparison of Turbine Housing Models 81

7   Application of the 0D-Turbocharger **89**

7.1   Introduction of the Engine and the Operating Point 89

7.2   Calibration of the Engine Models 93

7.3   Stationary Analysis: Part Load Operating Point 94

7.4   Transient Analysis: Load Change via Ignition Retarding 101

8   Conclusion and Outlook..**109**

Bibliography 113

Appendix...121

A.1   Appendix 1 121

Figures

2.1 H-S diagram for the compression and expansion process 14
2.2 Qualitative compressor performance map 18
2.3 P-V-diagrams of naturally aspirated and turbocharged engines 20
2.4 Spectrum of unsteady events according to [17] 22
2.5 Influence of amplitude on the level of unsteadiness [17] 24
3.1 Time scale of different simulation approaches [60] 30
3.2 Simulation domain increase for reduced dependency of boundary conditions [60] 31
4.1 Four-cylinder engine with boundary conditions [60] 37
4.2 Example pressure traces close to turbocharger 41
4.3 Approach 1 (left) and 2 (right) for the integrated turbocharger 44
4.4 Approach 3 (left) and 4 (right) for the integrated turbocharger 45
5.1 Concept QuickSim with 0D-Turbocharger 47
5.2 Flow chart of the integrated 0D-Turbocharger 48
5.3 Schematic for calculation of pressure recovery in diffuser 49
5.4 Implementation of surge model mass flow 52
5.5 Implementation of surge model efficiency 53
5.6 Turbine performance map with four quadrants adapted from [23] 54
5.7 Turbine performance map of the turbocharger used in chapter 6 55
6.1 1D-Simulation model of the virtual hot gas test bench 60
6.2 QuickSim model of the virtual hot gas test bench 60
6.3 3D-CFD model of the virtual hot gas test bench 61
6.4 Heat capacity versus temperature at constant pressure (1bar) 62
6.5 Positioning and nomenclature of the sensors 63
6.6 Artificial pressure pulses with five-cylinder example 64
6.7 Pressure and mass flow at system inlet for the 50 Hz pulse 67
6.8 Pressure and mass flow at system inlet for the 400 Hz pulse 68
6.9 Pressure and mass flow at housing inlet for the 50 Hz pulse 69
6.10 Pressure and mass flow at housing inlet for the 400 Hz pulse 70
6.11 Total pressure at turbine inlet for 50 Hz (left) and 400 Hz (right) 71
6.12 Torque at turbine inlet for 50 Hz (left) and 400 Hz (right) 72

6.13 Torque for 1, 2 and 3 bar amplitude and a 50 Hz pulse 74

6.14 Torque for 1, 2 and 3 bar amplitude and a 400 Hz pulse.................. 75

6.15 Torque trace for 50k, 100k and 150k rpm and a 50 Hz pulse............ 77

6.16 Torque trace for 50k, 100k and 150k rpm and a 400 Hz pulse 78

6.17 Influence of manifold length on mass flow trace 80

6.18 Influence of temporal discretization on a 400 Hz pressure pulse 81

6.19 PR sensor concepts for the turbine housing................................ 82

6.20 Definition of the volute dimensions in accordance with [11]............ 83

6.21 Different versions of the implemented 3D turbine housing 84

6.22 Torque trace for versions A - C at 400 Hz pressure pulse 85

6.23 Torque trace for Version C - F at 400 Hz pressure pulse................. 86

7.1   MPE-850 simulation domain in test bench setup [32].................... 90

7.2   Mesh representation (left) and porous representation (right) 91

7.3   Comparison of intake pressure for different throttle models............ 92

7.4   Valve lift of the two-cylinder engine....................................... 95

7.5   Pressure in the intake runner [32] 95

7.6   Pressure in exhaust runners [32].. 96

7.7   Reversed mass flow in turbine during one working cycle [32] 97

7.8   Turbocharger speed and rotor power during a working cycle [32] 98

7.9   Pressure traces with and without 0D-Turbocharger [32]................. 99

7.10 Mass flow with and without 0D-Turbocharger [32]......................100

7.11 Turbocharger speed after load change [32] 103

7.12 Turbine flow enthalpy for test bench shaft speed [32]....................104

7.13 Intake pressure during the transient maneuver 105

7.14 Exhaust pressure during the transient maneuver..........................106

7.15 Mass flow during the transient maneuver..................................106

A1.1 Calibration procedure for QuickSim with 0D-Turbocharger 123

Tables

4.1   Maximum requirements derived from figure 4.2 42

4.2   Estimated calculation time (per cycle) for different approaches 45

5.1   Friction constants for eq. 5.8 from [45]...................................... 57

6.1   Overview 1D, QuickSim and 3D-CFD simulation 62

6.2   Characteristic numbers of the artificial pressure pulses................... 64

6.3   Stationary calibration results of the VHGTB 65

6.4   Pressure difference in boundary cells of volute outlet..................... 87

7.1   Textron MPE 850 four-stroke engine specification 89

7.2   Operating point of the two-cylinder engine 92

7.3   Two-cylinder engine stationary results after the calibration process ... 93

7.4   Optimization of calculation time for one working cycle in full load .. 101

A1.1 Models chosen in STAR-CCM+ for the calculation of the aerodynamic turbine performance map 121

A1.2 Models chosen in STAR-CCM+ for the calculation of pressure pulses on the VHGTB 121

A1.3 Models chosen in GT-Power for the calculation of VHGTB and the two-cylinder engine 122

A1.4 Approaches for an integrated turbocharger 122

Abbreviations

0D	Zero-Dimensional
1D	One-Dimensional
3D	Three-Dimensional
BMEP	Break Mean Effective Pressure
CA	Crank Angle
CAD	Computer-Aided Design
CCL	Closed Compressor Loop
CFD	Computational Fluid Dynamics
CPU	Central Processing Unit
DNS	Direct Numeric Simulation
EGR	Exhaust Residual Gas
FBDC	Firing Bottom Dead Center
FKFS	Forschungsinstitut für Kraftfahrwesen und Fahrzeugmotoren Stuttgart
FTDC	Firing Top Dead Center
HIR	High Inertia Rotor
ICE	Internal Combustion Engine
IMEP	Indicated Mean Effective Pressure
IVK	Institut für Verbrennungsmotoren und Kraftfahrwesen
LES	Large Eddy Simulation
MF	Mass Flow

NVH	Noise, Vibration, Harshness
RANS	Reynolds-Averaged Navier-Stokes
RDE	Real Driving Emissions
SAE	Society of Automotive Engineers
SBDC	Scavenging Bottom Dead Center
SCR	Selective Catalyst Reaction
STDC	Scavenging Top Dead Center
VHGTB	Virtual Hot Gas Test Bench
VTG	Variable Turbine Geometry
WLTP	Worldwide harmonized Light vehicles Test Procedure
WOT	Wide Open Throttle

Symbols

Latin Letters

A	Area	m^2
c_f	Long-range process term in conservation equation	-
Co	Courant number	-
c_p	Specific heat at constant pressure	J/kgK
F	Extensive variable in conservation equation	-
f	Density / intensive variable in conservation equation	-
G	Gravitational energy	J/kg
H_E	Exhaust enthalpy	J
h_f	Heat of formation	J/kg
H_I	Intake enthalpy	J
H_L	Leakage enthalpy	J
h_{tc}	Thermo-chemical enthalpy	J/kg
h	Enthalpy	J/kg
i	Stroke factor	-
$\vec{j}$	Diffusive mass flux	kg/m^2s
K	Wrinkling factor	-
k	Turbulent kinetic energy	$m/^2/s/^2$
l	Characteristic length	m
M	Molar mass	kg/kmol
m	Mass	kg
m_B	Fuel mass	kg
m_C	Cylinder mass	kg
m_E	Exhaust gas mass	kg
m_I	Intake mass	kg
m_l	Leakage mass	kg
$\dot{m}$	Mass flow	kg/s
n	Engine speed	rpm

P	Power	W
p	Pressure	Pa
p_{me}	Mean effective pressure	Pa
$\overline{\overline{P}}$	Second order stress tensor	N/^{2}
PR	Pressure ratio	-
Q_B	Fuel heat release	J
Q_W	Wall heat transfer	J
R_s	Specific gas constant	J/kgK
Re	Reynolds number	-
S	Flame speed	m/s
s_f	Source or sink term in conservation equation	-
St	Strouhal number	-
T	Temperature	K
s	Entropy	J/K
t	Time	s
$\vec{V}$	Diffusive fluid velocity	m/s
V	Volume	m^3
$\vec{v}$	Fluid velocity	m/s
V_H	Engine displacement	m^3
W	Work	J
w	Mass fraction	-

Greek Letters

ε	Dissipation rate of k	m^2/s^3
η	Efficiency	-
Γ	Specific pressure gradient	bar/rev
Λ	Criterion for unsteadiness	-
γ	Ratio of specific heats	-
Δh	Difference between two enthaply levels of a process	J/kg
μ	Dynamic viscosity	kg/ms
ω_i	Molar formation rate of species i	kmol/m^{3}s
φ	Crank Angle	°CA
$\vec{\Phi}_f$	Flux of the intensive variable f in the conservation equation	-

$\overline{\overline{\Pi}}$	Second order shear stress tensor	N/m^2
ρ	Mass density	kg/m^3
ρ_i	Mass density of species i	kg/m^3

Indices

0	Characteristic	
1	Compressor inlet	
2	Compressor outlet	
3	Turbine onlet	
4	Turbine outlet	
22	Post charge air intercooler	
act	Real process	
amb	Ambient	
amp	Amplitude	
B	Bearing	
C	Compressor	
corr	Corrected	
D	Discretization	
e	Effective	
f	Flame	
frict	Friction	
i	Species i	
is	Isentropic process	
lam	Laminar	
mech	Mechanical	
oil	Properties of the used oil	
ref	Reference	
s	Static	
sh	Shaft	
T	Turbine	
t	Total	
tm	Thermo-mechanic	
trb	Turbine	
ts	Total-static	
tt	Total-total	

turb	Turbulent
vol	Volute
wheel	Turbo machinery rotor

Abstract

Driven by the competition for the powertrain of the future, the internal combustion engine research requires new technologies, like the virtual engine development, for further improvements while decreasing time and cost.

The present work aims to develop and validate an approach to allow 3D-CFD simulations of turbocharged combustion engines under stationary and transient conditions. To simulate an engine load change, the modeling has to be able to cover multiple physical seconds and more than 100 consecutive working cycles within an acceptable calculation time. The work presented here was carried out in the QuickSim framework, a 3D-CFD code dedicated to internal combustion engines developed at the FKFS/IVK in Stuttgart, Germany.

Numerous approaches, varying in simulation effort and depth, were compared in order to find the optimal one. The approach chosen, models the entire geometry (intake and exhaust system before and after the turbocharger, cylinders etc) three dimensional, with the exception of the rotor passages. Turbine and compressor rotors are replicated by a zero dimensional approach, enabling the use of widely available performance maps. To extend the operating range, additional models for reversed mass flow and compressor surge or choke are implemented.

Using the example of a turbocharger turbine, the modeling is validated on a virtual hot gas test bench for stationary and pulsed flow against 1D and 3D-CFD simulations, which are the widely recognized industry standard. Sinusoidal pressure pulses varying in amplitude and frequency are designed to replicate measured exhaust pulses. All three simulation environments are compared with respect to accuracy and stability, revealing significant differences between the two references for high pressure gradients. The results of the implemented modeling fall between the two references, showing good agreement for pulses with a medium pressure gradient and acceptable agreement for worst case scenarios.

The validated modeling is used to simulate the complex gas exchange process of a two-cylinder engine in stationary part load operation. Time resolved pressure traces from the intake and exhaust system are compared to measurements from the test bench, showing good agreement despite the difficult operating conditions. Simultaneously, the engine is used to demonstrate the superiority of the integrated turbocharger over a conventional time-dependent boundary condition with regard to accuracy and stability.

Starting from the stationary operating point, a change in load is initialized through a severe ignition retarding. The following transient, characterized by the increasing turbocharger speed, boost pressure etc, is compared to measurements from the test bench. It is demonstrated that a transient 3D-CFD simulation of a turbocharged engine is possible with a 3D-CFD based virtual engine development. At the same time, the necessity for advanced wall temperature modeling and controllers to replicate the test bench is highlighted.

Kurzfassung

Um im Wettbewerb für den Antriebstrang der Zukunft bestehen zu können, benötigen die Entwickler von Verbrennungsmotoren Werkzeuge wie die virtuelle Motorenentwicklung, um schnell und mit akzeptablen Kosten weitere technologische Verbesserungen voranzutreiben.

Die vorliegende Arbeit behandelt die Entwicklung und Validierung eines Simulationswerkzeugs zur Analyse aufgeladener Verbrennungsmotoren mit Hilfe moderner 3D-CFD Strömungssimulationen. Neben der Untersuchung stationärer Motorbetriebspunkte stehen insbesondere transiente Bedingungen, z.B. ein Lastwechsel, im Fokus, welche häufig mehr als 100 aufeinander folgende Arbeitsspiele umfassen. Die Herausforderung besteht, insbesondere für eine 3D-CFD Simulation, darin, dies in einer akzeptablen Rechenzeit zu bewerkstelligen. Die Arbeit wurde innerhalb der auf Verbrennungsmotoren spezialisierten Entwicklungsumgebung QuickSim durchgeführt, welche am FKFS/IVK in Stuttgart entwickelt wurde.

Mehrere Ansätze, die sich hinsichtlich der Simulationstiefe und des -aufwands unterscheiden, werden verglichen, um den am besten geeigneten zu identifizieren. Im gewählten Ansatz wird der komplette Luftpfad (Ansaug- und Abgassystem vor und nach dem Turbolader, Zylinder, etc.) - mit Ausnahme der Laufradpassage - dreidimensional dargestellt. Die Rotoren der Turbomaschinen werden durch kennfeldbasierte 0D-Modelle integriert. Um den Arbeitsbereich von Turbine und Verdichter an die Erfordernisse des verbrennungsmotorischen Betriebs anzupassen, werden zusätzliche Modelle für eine Umkehr des Massenstroms sowie der Pump- und Stopfgrenze implementiert.

Die Validierung der Modelle erfolgt mittels eines virtuellen Heißgasprüfstands, auf welchem diese unter stationären und gepulsten Bedingungen mit konventionellen 1D und 3D-CFD Strömungssimulationen verglichen werden. Dazu werden sinusförmige Druckverläufe definiert, deren Druckgradienten in Anlehnung an gemessene Werte gewählt werden. Die drei verglichenen Simulationsumgebungen zeigen deutliche Unterschiede, welche insbesondere zwischen den zwei gewählten konventionellen Referenzen, mit dem maximalen

Druckgradienten zunehmen. Die Ergebnisse der vorgestellten Implementierung liegen zwischen den Referenzen und zeigen eine gute Übereinstimmung für mittlere und eine akzeptable Übereinstimmung für extrem hohe Druckgradienten.

Die validierte Modellierung wird im Folgenden auf einen Zweizylindermotor angewendet, der im ausgewählten Teillast-Betriebspunkt einen herausfordernden Ladungswechsel aufweist. Die Messsignale der Niederdruckindizierung im stationären Betrieb zeigen trotz der schwierigen Randbedingungen eine sehr gute Übereinstimmung mit der Simulation. Gleichzeitig werden die Vorteile der Modellierung bezüglich Genauigkeit und Stabilität im Vergleich zu konventionellen, zeitaufgelösten Randbedingungen aufgezeigt.

Beginnend mit dem bereits bekannten Teillastpunkt wird ein Lastsprung des Motors durch eine deutliche Rücknahme des Zündwinkels ausgelöst. Der folgende transiente Motorbetrieb ist durch den Anstieg von Lade- und Abgasdruck sowie der Turboladerdrehzahl gekennzeichnet und wird mit den entsprechenden Messungen am Prüfstand verglichen. Hierdurch wird gezeigt, dass eine transiente Motorsimulation inklusive Turbolader mit 3D-CFD basierter virtueller Motorenentwicklung möglich ist. Gleichzeitig wird gezeigt, dass in diesem Fall eine adäquate Modellierung des Wandwärmeübergangs sowie eine exakte Abbildung der Prüfstandsregler benötigt werden.

1 Introduction

Over the course of history, increasing the mobility of humans and goods has been a major driver for social development. In the most recent decades, the internal combustion engine has secured a dominant position within the multitude of known drive concepts. Animated by the competition with alternative technologies, continuous improvements with well-known milestones like charged engines, lead free fuels, the 3-way catalyst and modern exhaust gas after-treatment have been introduced.

Until today, the requirements for a successful drive concept have not changed; however, the focus has been shifted towards a reduction of the environmental impact. In consequence, energy consumption during production and lifetime as well as local and global emissions have to be reduced. This is also reflected by the regulations for measuring both criteria: artificial tests have been reduced in favor of measurements on the roads, e.g., real driving emissions. Under these circumstances the transient engine behavior - over a wide range of operating conditions - has gained importance. While the transient behaviour of naturally aspirated engines is usually uncritical, supercharged and especially turbocharged engines show an increased complexity under transient loads, which has to be addressed during development. The most famous manifestation is probably the turbo lag, in which the spin up time before enough boost pressure is created causes a significant delay between the torque request from driver and the engine delivering.

Virtual engine development is a concept aiming to increase process speed and efficiency, while lowering cost. Despite its name, it does not rely solely on simulation technology; instead, the test bench and simulation complement each other. The individual work share is defined dependent on the project requirement with a tendency to reduce the number of measurements, and especially the different hardware variants, while at the same time increasing the variety and quality of measurement methods. Simulations are largely dependent on calibration as they either use empirical or simplified physical models. In both cases, the quality of the calibration data strongly determines the quality of the

© Springer Fachmedien Wiesbaden GmbH, part of Springer Nature 2020
A. Kächele, *Turbocharger Integration into Multidimensional Engine Simulations to Enable Transient Load Cases*, Wissenschaftliche Reihe Fahrzeugtechnik Universität Stuttgart, https://doi.org/10.1007/978-3-658-28786-3_1

simulation results. An iterative approach is used, with the simulation proposing new geometries or operation strategies, and the test bench validating or calibrating the simulation.

As of 2019, there is no industry standard for the use of 3D-CFD in internal combustion engines. The application ranges from simulations of the intake channels with cold and dry air, over single cycle models with or without combustion, to the advanced approaches capable of simulating a full engine for multiple consecutive working cycles. To diminish the impact of the boundary conditions on the results, the simulation domain is often extended to include the air box and sometimes even the entire air path of the car or the test bench.

These advanced simulation tools are capable of simulating a stationary point of operation with several consecutive working cycles to reach a converged state for the cycle-averaged values like IMEP, efficiency etc. If the engine is turbocharged, simulation complexity increases, as the flow boundary conditions are no longer environmental, but have to be either measured or known from another simulation. Both ways can deteriorate the simulation quality, due to measurement errors or the limitations of the previous simulation.

To ensure compliance with real driving emissions, the transient engine behavior is of remarkable importance. Ironically, this is where the 'fixed' boundary conditions from an external source are no longer usable with turbocharged engines. By predetermining the boundary conditions, the transient interaction between turbocharger and engine is missing entirely. A possible solution is the introduction of the turbocharger into the engine simulation, allowing environmental conditions before the compressor and after the turbocharger again.

Generation of emissions, as a product of the combustion, is a highly three-dimensional process underlining the need for a 3D-CFD simulation to capture and fully understand the phenomena. An emission prediction for a drive profile is commonly performed with 1D-Simulations calibrated to measurements due to the computation time.

The ideal solution and, consequently, the goal of the present work, is a 3D-CFD simulation that integrates the turbocharger fully into the simulation of the combustion engine and can be used not only for stationary operating conditions but also transients, like a load change lasting typically for multiple seconds and easily including one hundred or more working cycles.

In Chapter 2, the fundamentals of downsized automobile consumer engines and the real working process analysis are introduced. The methodology of 1D and 3D-Simulations is discussed with focus on the 3D-CFD. Different efficiency definitions for turbochargers and the turbomachines included are presented in conjunction with characterizing numbers. Multiple ways to describe the level of unsteadiness of pressure pulses are shown by different characteristic numbers.

The three simulation environments used throughout this work are presented in Chapter 3, with priority given to the 3D-CFD code QuickSim. This is continued in Chapter 4, where the state of the art and different approaches for the integrated turbocharger are discussed.

Chapter 5 is dedicated to the chosen approach of a performance-map-based 0D-Turbocharger. Here the general modeling of the turbocharger is introduced as well as additional models to widen the operating range outside of the first quadrant.

The modeling is validated through a virtual hot gas test bench in Chapter 6. Artificial pressure pulses are derived from the exhaust pressures of real engines and a worst case scenario is defined. Frequency, amplitude and turbocharger speed are varied, and the results compared to a 1D and a 3D-CFD environment. Finally, different options for the turbine housing are investigated.

In Chapter 7, the validated modeling is applied to a two-cylinder engine. To demonstrate the stability, a difficult part load operation point has been chosen to be investigated under stationary conditions. In a second step, a change in engine load is simulated through ignition retarding for more than 100 consecutive working cycles.

2 Fundamentals

The modern internal combustion engine (of the year 2019) for passenger cars is no longer naturally aspirated, but uses charged air instead. This is usually realized with a turbocharger, less often with a supercharger and very rarely with an electric charger. The percentage of new cars sold with a turbocharger is rising in all major markets with a positive outlook [53]. Several derivatives of these systems, like multiple chargers in serial and parallel connection or turbo compound systems, have been developed. However, due to the increasing complexity and additional cost, these systems are only seen in high-end consumer cars or heavy trucks. The average consumer car has only one turbocharger to decrease the engine size, leading to the well-known downsized engine. The engine speed can also be lowered to reduce friction (downspeeding).

The power of any internal combustion engine can be calculated using equation eq. 2.1. With the effective power P_e , the factor i for two or four stroke engines, the break mean effective pressure (BMEP) p_{me} and the displacement of the engine V_H.

$$P_e = i * n * p_{me} * V_H \qquad \text{eq. 2.1}$$

In a downsized or downspeeded engine, V_H or n are reduced and therefore p_{me} has to be increased in order to maintain the same peak power. By increasing the specific load, the engine is able to run more efficiently as the gas exchange and specific heat losses are reduced [24]. At the same time, the increased power density of a downsized engine adds design challenges requiring advanced solutions. The injection and charge system, as well as the overall mechanical and tribological capabilities of the engine, are especially important. The theoretical efficiency benefit can not always be fully exploited due to limiting factors like the required enrichment under full load condition.

© Springer Fachmedien Wiesbaden GmbH, part of Springer Nature 2020
A. Kächele, *Turbocharger Integration into Multidimensional Engine Simulations to Enable Transient Load Cases*, Wissenschaftliche Reihe Fahrzeugtechnik Universität Stuttgart, https://doi.org/10.1007/978-3-658-28786-3_2

2.1 The Real Working Process Analysis

In order to comprehend and analyze the in-cylinder thermodynamic processes, the real working process analysis has been developed.It is based on the mass and energy balance of the cylinder and derived from equation eq. 2.2 and eq. 2.3.

$$\frac{dm_C}{d\varphi} = \frac{dm_I}{d\varphi} + \frac{dm_E}{d\varphi} + \frac{dm_L}{d\varphi} + [\frac{dm_B}{d\varphi}]_{DI} \qquad \text{eq. 2.2}$$

The change of cylinder mass m_C is the result of exhaust gas m_E and leakage mass m_L, due to blow-by, leaving the cylinder as well as the inflow of fresh load m_I and for a direct injection the fuel mass m_B.

$$Q_B + Q_W + H_I + H_E + W + H_L = 0 \qquad \text{eq. 2.3}$$

The conservation of energy can be written as the fuel heat release energy Q_B, wall heat transfer Q_W, intake enthalpy H_I, exhaust enthalpy H_E, leakage enthalpy H_L and work W. The differential notation of eq. 2.3 describes the change of internal energy U over time and is referenced to the common notation for the engine operating cycle φ:

$$\frac{dQ_B}{d\varphi} + \frac{dQ_W}{d\varphi} + \frac{dH_I}{d\varphi} + \frac{dH_E}{d\varphi} + \frac{dW}{d\varphi} + \frac{dH_L}{d\varphi} = \frac{dU}{d\varphi} \qquad \text{eq. 2.4}$$

Using the ideal gas equation in eq. 2.5 the relation between the thermal state variables, namely pressure p, volume V, mass m, specific gas constant R_s and temperature T can be described.

$$p \cdot V = m \cdot R_s \cdot T \qquad \text{eq. 2.5}$$

The real working process can be used with different approaches, i.e. one-zone, two-zone or n-zone. Depending on the combustion process, different approaches are favored, e.g. two-zone for flame propagation or n-zone for diesel combustion. The mixture is assumed to be perfectly homogenized, concerning temperature, pressure and species concentration. After the ignition, an infinitesimal small flame front is assumed, dividing the two zones in a burned and unburned region. With the burned region growing spherically, the flame front moves through the combustion chamber. To determine the combustion rate, a model is implemented e.g. the vibe function, however these models need to be calibrated and are not predictive.

Today, the real working process is mainly used on the test bench in order to evaluate the combustion in real time ('online') from measurable quantities like the cylinder pressure trace. For predictive simulations, or even engine development, other simulation tools offer benefits like spatial resolution inside the cylinder, or the extension of the simulated domain to the intake and exhaust system.[4, 15, 37]

2.2 1D-Simulation

Historically, 1D-Simulations are an extension of the real working process analysis, delivering information about the cylinder gas exchange and the trapped air/mixture mass. This is done by extending the simulation domain to parts of the intake and exhaust region by adding single [4] or a number of discretized volumes before the intake and after the exhaust valves [63]. With further development, the complete flow path through the engine can be modeled by means of pipes (including options for tapered sections and bends), flow splits, and special models like turbomachines, sensors etc. This allows a one dimensional spatial resolution in addition to the timely resolution.

Several sub-models have been developed to adapt the 1D-Simulation methodology to the needs of the internal combustion engine, e.g. cylinder heat transfer, cranktrain friction or temperature shift of thermocouples. A number of approaches have been proposed, aiming to replace the fixed combustion profiles of the real working process analysis with empirical or quasi-dimensional models to gain better predictive power and reduce the calibration influence [25, 43, 46].

For each of the described discretized flow elements, the continuity, momentum and energy equation have to be solved in every time step or iteration [15]. The required simulation time is still low but usually too high for online calculation on the test bench. Through the very comprehensive results and the widely available commercially tools, 1D-Simulations are nowadays mainly used in the engine or automobile development. With the integration of the oil and water cooling circle, oder a camshaft design option, a broad range of tasks can be handled in one simulation environment conveniently. The drawback of 1D-Simulation is by nature the lack three dimensional information, which is essential for some of the processes in a combustion engine like fuel and water injection, mixture formation or scavenging. Due to the lack of this information and other simplifications made, intensive calibration to accurate test bench data is required. The fields of use are often:

- Investigation of charging concepts / turbocharger matching

- Intake and exhaust manifold layout

- Engine transient behavior

- Valve timing design

- Vehicle / track simulation

2.3 3D-CFD Simulation

To make a three-dimensional simulation domain of any given shape and size accessible for simulation, the continuum is discretized into a finite number of cells, in order to reproduce the desired geometry. Different cell shapes can be choose: the most popular are tetrahedral, hexahedral or octahedral. A center is defined to represent the fluid properties and dynamic state of each cell with the assumption that both are homogeneous within one cell (finite volume approximation). Time is discretized similarly into time steps of an appropriate size: in the case of a transient flow, the solution of the time step t_{n+1} is derived from the time step t_n. As the fundamentals of computation fluid dynamics are well established the reader may refer to [1, 33, 37, 39, 58]. A brief overview based on [15] will be given in the following paragraphs.

2.3.1 Fundamental Equations

The fundamental equations describe the conservation of extensive variables that can be derived from the Euler formulation in eq. 2.6 with the corresponding variable density or intensive variable $f(\vec{x},t) = dF/dV$.

$$\frac{\partial f}{\partial t} + div\left(\vec{\Phi}_f\right) = s_f + c_f \qquad \text{eq. 2.6}$$

The equation states that a change in the variable density $\partial f/\partial t$ can be caused by a flux through the volume surface $\vec{\Phi}_f$, by a source or sink (s_f) or a long-range process (c_f) like radiation or gravitation. The following conservation equations for mass, momentum and energy can be directly derived from this equation.

2.3.2 Mass Conservation

Adapting eq. 2.6 to the case of mass conservation, the extensive variable $F(t)$ is replaced by the mass and the density variable $f(\vec{x},t)$ by the mass density ρ. Flux density can be written as the product of local flow velocity $(\vec{v})$ and *rho*. The source or sink therm s_f as well as the long range therm are set to zero, as mass cannot be created or destroyed by the processes investigated.

$$\frac{\partial \rho}{\partial t} + div(\rho\vec{v}) = 0 \qquad \text{eq. 2.7}$$

This can be extended to multiple species, for example air and fuel vapor, by replacing the mass (m) by the mass fraction $w_i = m_i/m$ of species i. Mass density can be written as the product of local flow velocity v_i composed of the mean flow velocity $\vec{v}$ and the diffusion velocity $\vec{V}_i$. The diffusion mass flux created by the diffusion velocity can be written as $\vec{j}_i$ as displayed in eq. 2.8.

$$\vec{\Phi}_f = \rho_i \vec{v}_i = \rho_i\left(\vec{v} + \vec{V}_i\right) = \rho_i\vec{v} + \vec{j}_i \qquad \text{eq. 2.8}$$

As species can be transformed into each other, e.g. through combustion or other chemical reactions, a source term is required which can be written as the product of species molar mass M_i and species molar formation rate ω_i.

$$\frac{\partial \rho w_i}{\partial t} + div\,(\rho w_i \vec{v}) + div\,\vec{j_i} = M_i \omega_i \qquad \text{eq. 2.9}$$

2.3.3 Momentum Conservation

For the conservation of momentum, the density variable is replaced by the momentum density $(\rho \vec{v})$, the flux is represented by a convective part $\rho \vec{v} \otimes \vec{v}$ and the second order stress tensor $\overline{\overline{P}}$ which is defined by the momentum change due to viscous effects $\overline{\overline{\Pi}}$ (shear stress tensor) and the pressure p.

$$\vec{\Phi}_f = \rho \vec{v} \otimes \vec{v} + \overline{\overline{P}} = \rho \vec{v} \otimes \vec{v} + p\overline{\overline{I}} + \overline{\overline{\Pi}} \qquad \text{eq. 2.10}$$

The momentum equation also requires a long range terms $(\rho \vec{g})$ to account for gravitation.

$$\frac{\partial\,(\rho \vec{v})}{\partial t} + div\,(\rho \vec{v} \otimes \vec{v}) + div\,\overline{\overline{\Pi}} - grad\,p = \rho \vec{g} \qquad \text{eq. 2.11}$$

2.3.4 Energy Conservation

The contained energy can be written as the sum of the internal energy u, the kinetic energy $\frac{1}{2}|\vec{v}|^2$, the potential gravitational energy G and the heat of formation of the contained species h_f.

$$\rho e = \rho \left(u + \frac{1}{2}|\vec{v}|^2 + G + h_f \right) \qquad \text{eq. 2.12}$$

The energy flux can be written as the sum of a convective term $\rho e \vec{v}$, the energy transport due to pressure and shear stress $\overline{\overline{P}}\vec{v}$ and the energy transport $\vec{j_q}$ e.g. due to heat conduction. As energy cannot be created or destroyed by engine processes, the source term s_f can be set to zero. The long range term c_f is required to take magnetic fields and radiation into account, represented by q_r.

The thermo-chemical enthalpy (h_{tc})itself can be described as the sum of the thermal enthalpy h and the heat of formation h_f.

$$\frac{\partial (\rho h_{tc})}{\partial t} - \frac{\partial p}{\partial t} + div\left(\rho h_{tc}\vec{v} + \vec{j}_q\right) + \overline{\overline{P}} : grad(\vec{v}) - div(p\vec{v}) = q_r \qquad \text{eq. 2.13}$$

2.3.5 Turbulence

In general, literature describes two flow regimes: the laminar flow, in which the vectorial and scalar values are regularly dispersed and the flow can be described as a pack of fibers, and the turbulent flow, in which chaotic instationary structures of different size exist. The reason for this chaotic flow can be found in the shear stresses, creating vortices of different size and shape. They are continuously drawing energy from the main flow, while shedding into an increasing number of smaller vertices. During this vortice cascade, energy is dissipated back to the main flow. A way to characterize the dominant regime is the Reynolds number in equation eq. 2.14, describing the ratio of inertial to viscous forces.

$$Re = \frac{\rho |\vec{v}| l}{\mu} \qquad \text{eq. 2.14}$$

The inertial force is composed of the density ρ, the velocity $|\vec{v}|$ and the characteristic length l, while the viscous force is represented by the dynamic viscosity μ. Reynolds numbers, describing the transition between laminar and turbulent flow, have been found experimentally for specific cases. For example the flow through a smooth pipe with the characteristic length of the pipe diameter has been found to have the transition from laminar to turbulent flow at a Reynolds number of approx. 2300. Above this value, the turbulent phenomena (strong mixing, characteristic energy dissipation) will continuously increase.

Turbulence Modeling

In order to detect and calculate a vortice in 3D-CFD, the discretization length or cell size has to be chosen small enough for the vortice to be covered by at least two cells. A small cell size leads to a correspondingly fine timely discretization increasing the computational effort remarkable (see Courant-Friedrichs-Lewy Criteria or Courant number). The approach to calculate the entire flow field, including the smallest vortices, is known as Direct Numeric Simulation (DNS) and is only used for very specific applications as the computational effort is enormous. In a flow field, the size of the smallest vortice is directly related to the Reynolds number (see Kolmogorov length), therefore, the effort increases remarkably with increasing Reynolds number [33].

To reduce this effort, several classes of turbulence models have been introduced, differing mainly in the modeling depth. Large Eddy Simulations (LES) will only calculate the vortices larger than the cell size, while the smaller ones are modeled in a subgrid model. The larger vortices contain most of the vortice kinetic energy, rendering the loss in simulation depth and information acceptable for most applications. For the necessities of an iterative design process, however, the computational effort is still too high (as of 2019) and the approach is mostly used in basic research or for turbulence model development and calibration.

The most commonly used class of models are the Reynolds-Averaged Navier-Stokes models (RANS). They utilize turbulence models to describe the influence of turbulent fluctuations on the main flow, but not the fluctuations themselves. Additional equations, which are not based on exact physics, but rather simplified assumptions and empirical data, are required to obtain a closed system. A multitude of models have been developed for different applications and described in literature [21, 33, 39, 55, 59].

One of the best known representatives for RANS is the k-ε-model, with k being the turbulent kinetic energy and ε being the dissipation rate of k. The equations can be derived from the transport equations and optimized using model constant for different applications. The approach is well tested and several correction formulations have been found to account for unique flow phenomena, especially in the cases of a flow close to a stationary wall with high positive pressure gradients or other not entirely turbulent flow phenomena with high viscous influence.

Turbulence in the Internal Combustion Engines

The internal combustion engine is strongly influenced by turbulence (for example, the combustion processes, mixture formation or heat transfer) [15] and therefore by the turbulence model used. However, it is impossible to measure the turbulence inside a combustion chamber during operation, so no experimental data can be provided to calibrate or even choose a suitable model. To overcome this dilemma, one of many approaches is to use a well-known turbulence model and calibrate the secondary models (for example, the combustion or heat transfer model) to the test bench, as done in [15].

2.4 Turbocharger

2.4.1 Turbine and Compressor Characterization

A turbocharger is the combination of one or more compressors and one or more turbine wheels coupled via a common shaft. The turbine expands the exhaust gases, extracting power from the fluid, which is used to compress the charge air.

The pressure ratio (PR) for a compressor is defined as the ratio between the total pressure at the outlet and the total pressure at the inlet, as shown in equation eq. 2.15.

$$PR_C = \frac{p_{t,2}}{p_{t,1}}$$ eq. 2.15

In contrast to the compressor, the turbine pressure ratio is defined as the total pressure at the turbine inlet to the static outlet pressure. This is done to account for the fact that the kinetic energy leaving the turbine is not used and therefore lost to the process, while the kinetic energy leaving the compressor can be converted back into static pressure through the airbox volume.

$$PR_T = \frac{p_{t,3}}{p_{s,4}} \qquad\qquad \text{eq. 2.16}$$

Enthalpy flow and efficiency of the compression and expansion process can be visualized through the H-S diagrams depicted in Figure 2.1. Each of them contains four isobars for total and static pressure levels of the state before and after the rotor.

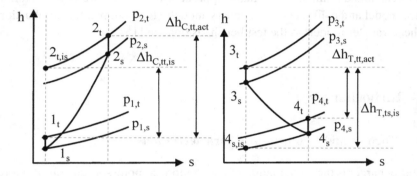

Figure 2.1: H-S diagram for the compression and expansion process

The compression process (shown on the left side) increases the total pressure, composed of static pressure and the velocity, written as specific pressure energy $\frac{\rho v^2}{2}$, from $p_{1,t}$ to $p_{2,t}$. Assuming an isentropic process $1_t 2_{t,is}$, no entropy is created, and flow enthalpy $\Delta h_{C,tt,is}$ has to be invested in order to achieve the pressure increase. The real process $1_t 2_t$, however, creates entropy and requires the significantly greater flow $\Delta h_{C,tt,act}$ of enthalpy to obtain the same $p_{2,t}$.

In the expansion process (see right side) which starts from $p_{3,t}$, the maximum achievable enthalpy extraction is $\Delta h_{T,ts,is}$ under the assumption that the kinetic energy can be reduced to zero. In a real turbine, the energy obtained from this process is less, as it is not an isentropic process and as the kinetic energy remains above zero.

The efficiency of both processes is defined as ratio between the actual work and a comparison process, namely an isentropic, politropic or isothermal process.

While there are applications where other definitions might be useful (e.g. poly-tropic efficiency for multistage turbomachines), in this work, efficiency will always be calculated in reference to the isentropic process. The compressor efficiency is defined as:

$$
\begin{aligned}
\eta_{C,tt,is} &= \frac{\text{isentropic compressor work}}{\text{actual compressor work}} \\
&= \frac{\Delta h_{C,tt,is}}{\Delta h_{C,tt,act}} = \frac{h_{2,t,is} - h_{1,t}}{h_{2,t} - h_{1,t}} \\
&= \frac{\frac{\gamma}{\gamma-1} R_s T_{1,t}}{c_p (T_{2,t} - T_{1,t})} \left[\left(\frac{p_{2,t}}{p_{1,t}} \right)^{\frac{\gamma-1}{\gamma}} - 1 \right]
\end{aligned}
\qquad \text{eq. 2.17}
$$

With the assumption of an ideal gas, equation eq. 2.17 can be written as

$$
\eta_{C,tt,is} = \frac{T_{1,t} \left(PR_{C,tt}^{\frac{\gamma-1}{\gamma}} - 1 \right)}{T_{2,t} - T_{1,t}}
\qquad \text{eq. 2.18}
$$

The turbine efficiency is defined as

$$
\begin{aligned}
\eta_{T,ts,is} &= \frac{\text{actual turbine work}}{\text{isentropic turbine work}} \\
&= \frac{\Delta h_{T,tt,act}}{\Delta h_{T,ts,is}} = \frac{h_{3,t} - h_{4,t}}{h_{3,t} - h_{4,s,is}} \\
&= \frac{c_p (T_{4,t} - T_{3,t})}{\frac{\gamma}{\gamma-1} R_s T_{3,t} \left[\left(\frac{p_{4,s}}{p_{3,t}} \right)^{\frac{\gamma-1}{\gamma}} - 1 \right]}
\end{aligned}
\qquad \text{eq. 2.19}
$$

With the assumption of an ideal gas, equation eq. 2.19 can be written as

$$
\eta_{T,ts,is} = \frac{T_{3,t} - T_{4,t}}{T_{3,t} \left(1 - \left(\frac{1}{PR_{T,ts}} \right)^{\frac{\gamma-1}{\gamma}} \right)}
\qquad \text{eq. 2.20}
$$

For the compressor, eq. 2.18 can be used to calculate the isentropic efficiency from measured temperatures and pressures. Due to the high temperature differences, resulting in high heat transfer from the exhaust gas to the turbine housing [35] and the strong vortex at the turbine outlet, this is not possible for the turbine. The efficiencies given in equation eq. 2.18 and eq. 2.20 are known as aerodynamic efficiencies, and characterize the aerodynamic properties of a turbomachine very well.

To overcome the measurement problems, a more convenient definition of the turbine efficiency, based on the energy conservation, can be used instead. The actual turbine power and the power loss due to friction are replaced by the actual compressor power. By doing so, the mechanical losses in the radial and axial bearings of the turbocharger shaft are added to the aerodynamic efficiency, resulting in the so called thermo-mechanical turbine efficiency.

$$\eta_{T,tt,is,tm} = \frac{P_{T,act} + P_{frict}}{P_{T,is}} = \frac{P_{C,act}}{P_{T,is}}$$

$$= \frac{\dot{m}_1 c_{p,1} (T_{2,t} - T_{1,t})}{\dot{m}_3 c_{p,3} T_{3,t} \left[1 - \left(\frac{1}{PR_{T,ts}} \right)^{\frac{\gamma-1}{\gamma}} \right]}$$

$$= \eta_{T,ts,is} \eta_{mech} \qquad \text{eq. 2.21}$$

Reduced and Corrected Characteristic Numbers

The compressor behavior is most often described by corrected parameters as defined below. Through the use of reference conditions (in this work 100 kPa and 293 K) the compressor behavior can be described independently of the inlet conditions. Corrected characteristic numbers maintain their unit and are therefore very comprehensive.

$$n_{C,corr} = n \sqrt{\frac{T_{ref}}{T_{t,1}}} \qquad \text{eq. 2.22}$$

$$\dot{m}_{C,corr} = \dot{m}\frac{p_{ref}}{p_{t,1}}\sqrt{\frac{T_{t,1}}{T_{ref}}}$$ eq. 2.23

The turbine behavior is rarely described by corrected values but almost exclusively by reduced speed and mass flow, because the inlet conditions can vary drastically. This is convenient as it makes comparison between two different machines run under different conditions possible. In the reduced form, however, the natural units are lost.

$$n_{T,red} = \frac{n}{\sqrt{T_{t,3}}}$$ eq. 2.24

$$\dot{m}_{T,red} = \dot{m}\frac{\sqrt{T_{t,3}}}{p_{t,3}}$$ eq. 2.25

To compare the behaviour of different gases, an additional term for the correction of the gas constant can be added.

Performance Maps

To characterize a turbine or compressor stage, performance maps measured under defined steady state conditions are the most established approach, as displayed in figure 2.2. They contain the mass flow and efficiency of each rotor depending on the shaft speed and pressure ratio and reflect special phenomena like surge or choke lines .

For the compressor and turbine test procedure, multiple institutions have published different standards like [52],[30] and [51]. According to the SAE standard the turbine inlet conditions are chosen at 600° C [49], thereby restricting the measurable operating range. On conventional hot gas test benches, the measurable area in the performance map of the turbine is further limited by the operating range of the compressor, namely by the surge and choke line. Several different methods have been developed to widen this range and cover the operation of ICEs. To name just a few: replacement of the compressor with an electric motor or a hydraulic brake [62], the closed compressor loop (CCL) [41], the high inertia rotor (HIR)[36] and reverse flow measurements [36].

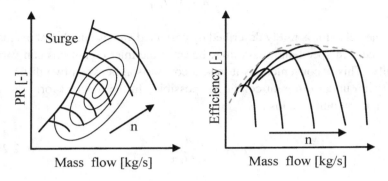

Figure 2.2: Qualitative compressor performance map

This additional information, however, is rarely provided by the manufacturer. To avoid these measurements, several different extrapolation methods have been developed. A common one has been presented by Jensen and Kristen-sens [31] and intensively tested and validated by others (e.g. [20]). A wide performance map is especially important for the turbine as the operating conditions can vary strongly during one engine working cycle, especially for pulse charging systems [48].

The efficiency measured on the test bench is not the aerodynamic, but the thermo-mechanical one (for conversion see equation 2.4.1) with two dominating phenomena: mechanical losses and heat transfer.

Mechanical losses in the turbocharger are dominated by the bearings, supporting the shaft both axially (thrust bearing) and radially (journal bearing). In automotive-size turbochargers, the most common type are hydrodynamic bearings, with either a fixed sleeve for low cost, or fully floating bearings with higher resilience and damping capabilities as well as reduced friction. Friction losses are especially important at low part load in order to maintain the desired boost pressure or to accelerate during a maneuver. Under these conditions, the exhaust enthalpy is low and the proportion of friction losses is high compared to the enthalpy available for expansion.

With the help of a friction test bench, the aerodynamic efficiency of the turbine stage can be reverse calculated from the thermo-mechanical efficiency.

$$\eta_{T,ts,is} = \frac{P_C + P_{frict}}{\dot{m}_T * (h_{3,t} - h_{4,s,is})} \qquad \text{eq. 2.26}$$

The second influencing phenomena is the internal heat transfer. During operation, the heat transfer from the hot turbine side to the colder compressor side through the shaft and common housing increases the compressor outlet temperature. With the compressor efficiency derived from the temperature and pressure increase, a higher heat transfer reduces the compressor efficiency noticeably in the performance map. This is especially true for low turbine speeds as the heat transfer is large compared to the work of the compressor. The turbine efficiency on the other side is measured too high, as lower compressor efficiency increases the power demand. [10]

The described heat transfer is especially critical for an oil-cooled turbocharger, as the oil flow is rather low and not designed to carry a lot of heat. In contrast, a water cooled turbocharger offers the water as a barrier between the hot and the cold side, reducing the mutual influences and referencing both sides to the water temperature [36].

To avoid the shortfalls of the thermo-mechanical performance map, 3D-CFD simulations are used to calculate aerodynamic performance maps. The compressor and turbine wheel can be simulated separately, increasing the measurable range and decreasing the influence of extrapolation. These maps do not include shaft friction nor heat transfer, requiring additional models with the ability for calibration. The separation between the performance map and heat transfer or friction is a great advantage for transient analysis, as both phenomena can change under transient conditions (e.g. warming of the turbine housing, etc).

2.4.2 Turbocharged Internal Combustion Engines

The turbocharger can be understood as an extension of the internal combustion engine as it pre-compresses air and post-expands exhaust gases. The compression of the intake air increases its density and, consequently, the air mass in the cylinders. At the same time, the remaining pressure difference between cylinder exhaust port and environment can be partially used, reducing the overall expansion losses and leading to a higher efficiency. The P-V diagram of a naturally aspirated engine (left side) and a turbocharged engine (right side) are depicted in Figure 2.3. The turbocharged engine operates with a significantly increased intake pressure (p_{22} vs. p_{amb}) and can even achieve a (partially) positive charge cycle.

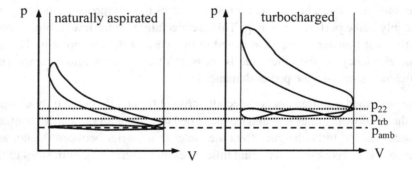

Figure 2.3: P-V-diagrams of naturally aspirated and turbocharged engines

By splitting the total compression work of the air charge between the compressor and cylinder, an intermediate cooling process is possible, allowing the temperature to be reduced and entropy to be removed, aiding compression efficiency (see divergence of isobars in the H-S diagram). In an internal combustion engine, the charge air temperature is especially important, as it largely determines the gas temperature of the unburned gas during combustion. A reduced charge temperature, therefore, allows higher compression ratios and/or an earlier center of combustion with identical knock limitations.

Mass flow through the internal combustion engine is highly scalable with the engine speed, which means reasonable efficiencies can be obtained over a wide power range. Turbomachines are generally more restricted in terms of mass flow range. As a very low mass flow is approached, the efficiencies drop due to very bad flow alignment to the rotor blade angles and, at high mass flow, the limited section area imposes a choking line where sonic speed is reached.

The transient response of a turbocharger is determined by the spin-up time to increase speed and boost pressure and can be represented through the rotor diameter. While a small turbine rotor might be desirable from the transient point of view, it strongly restricts the maximum flow capacity. Therefore, most of today's automobile turbocharges have a waste gate to bypass surplus exhaust gases around the turbine and allow for higher mass flow through the engine. More costly solutions to this problem are multiple turbocharges or a variable turbine geometry.

Turbochargers that are connected to reciprocating engines are subject to pulsating flow, created by the opening and closing of the exhaust valves. The pressure and temperature pulses at the turbine vary in shape, intensity and frequency, mainly influenced by the number of cylinders, the load case and the exhaust manifold configuration. Two extreme forms of flow motion have been reported on: constant pressure and pulse charging. Both of them are theoretical constructs describing the turbine inlet conditions, whereas reality is always a mix of both forms. Constant pressure charge systems are equipped with a large volume to dampen the pressure pulses emerging from the exhaust valve opening. The turbine wheel is designed for a high efficiency in stationary operation. In this method, not all of the 'blow-down' energy can be recovered from the exhaust gas, as the mixing process in the large volume cannot convert all of the kinetic energy in the pulse to the equivalent pressure. Pulse charging, on the other hand, requires a short distance between the exhaust valves and the turbine. It utilizes the kinetic energy of the flow in addition to the expansion and is attributed a better transient behavior [28]. In this case, the turbine requires high efficiencies, especially at a low ratio between circumferential speed and main flow velocity [22, 34].

For an automobile application where large volumes cannot be used due to packaging constraints, and where transient behavior is very critical, the turbine usually operates under pulse charging conditions. A higher number of cylinders per turbine results in an exhaust gas flow closer to constant pressure charging.

For multi-cylinder engines, literature states that 3 cylinders per turbocharger is the optimum compromise between good energy recovery and transient behaviour [65]. The operating conditions of the compressor are, in general, of a more constant nature as the volume between the compressor and the cylinders is larger and the amplitude of the pressure wave emerging from the opening of the intake valves is smaller.

2.4.3 Turbocharger under Transient Conditions

A turbocharger turbine in an automotive applications is affected by a large frequency spectrum of unsteady events with different amplitudes. Figure 2.4 displays this spectrum and the corresponding amplitude as proposed by [17]. The variations introduced by a transient engine maneuver is on the order of 1 Hz, while the exhaust pulse frequency is predicted between 10 and 100 Hz. Turbine blade passing frequency and turbulent fluctuations show significantly higher values.

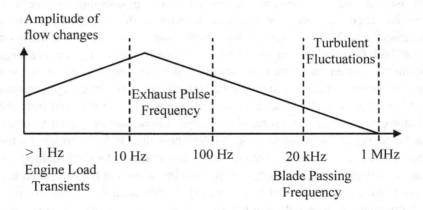

Figure 2.4: Spectrum of unsteady events according to [17]

Both extreme points -very low and very high frequencies- can be treated in a steady state manner. During engine transients, the time associated with the change is significantly longer than the time required by the flow to pass through the turbine; consequently, the behavior can be viewed as quasi-steady. For very high frequencies, the values can be averaged and treated as steady because the time scales of the fluctuations are multiple orders of magnitude lower than the convective flow and the amplitude of the fluctuations is very low

In contrast to these extreme ends, [17] expects the exhaust pulses to fall between 10 - 100 Hz (actually they can be much higher, as shown in Chapter 6.2) and show the highest of all amplitudes. It is therefore uncertain whether the exhaust pulses can be treated as quasi-steady in the simulation.

Most turbocharger measurements are done under steady state conditions for the sake of accuracy, the performance map thus obtained is, consequently, for the time being, only valid for steady state conditions. Many authors have published focusing on the application of these maps to non-stationary conditions [6, 8, 17, 18, 44, 56]. All authors agree that the non-stationary conditions can be divided in two regimes. Within the quasi-steady regime, the turbine behaves during each time step as if the conditions were constant. This has been found to be true for the rotor passage itself for most applications. However, the large volume of the turbine housing does show a remarkable filling and emptying behavior, leading to a strong deviation from the stationary map [38, 57]. If this characteristic is encountered through additional modeling (usually by a 0D or 1D approach), the turbine stage, comprised of the rotor and the turbine housing, can be treated as quasi-steady and the stationary performance map can be used [3, 12, 13, 14, 36, 42]. A common approach for 1D-Simulations models the turbine housing as a pipe with a length that is determined by the length of the housing inlet and half of its spiral part (e.g. [2]).

In the second regime, the quasi-steady assumption is not valid, resulting in a deviation from the performance map. Characteristic transient numbers have been introduced to characterize which regime is present.

One way to describe the level of unsteadiness for flow phenomena is the Strouhal number, giving the ratio between fluid travel time for a certain distance and the oscillation time of pulsations as displayed in equation eq. 2.27. With l_0 describing a characteristic length (e.g. length of a pipe or turbine housing), v_0 describing the fluid velocity and f_0 being the oscillation frequency. The frequency f_0 can be written as the reciprocal value of t_0 and the travel time T_0 can be derived from L_0 and v_0.

$$St = \frac{l_0 f_0}{v_0} = \frac{l_0}{v_0 t_0} = \frac{T_0}{t_0} \qquad \text{eq. 2.27}$$

For a Strouhal number significantly smaller than unity, the oscillation time-scale is remarkably lower than the fluid travel time. Therefore, the oscillations will not influence the mean value and the flow can be viewed as steady. However, with a Strouhal number approaching unity, the flow becomes unsteady and the oscillation has a great impact on the flow behavior. The Strouhal number was employed, for example, by Chen and Winterbone [13] with the findings that the turbine housing can reach high values of 0.1 while the runner itself remains relatively low at about 0.01. Further development has been carried out by Szymko et al. [50] to account for the fact, that the wave length is usually longer than the pulse length through the definition of a pulse length fraction factor. Other authors like Rajoo [42] and Costall [18] use a reduced frequency.

To consider not only the frequency but also the amplitude when determining the level of unsteadiness, Copeland et al. [17] introduced the lambda criterion. The influence of amplitude is displayed in Figure 2.5 below.

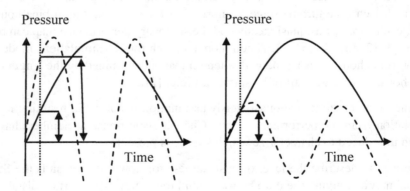

Figure 2.5: Influence of amplitude on the level of unsteadiness [17]

On the left side, the pressure amplitude is the same for both pulses, with obviously different pressure levels obtained after travel time (vertical dotted line). The right side shows a reduced amplitude for the higher frequency, leading to the same pressure. The unsteadiness of a flow has to take frequency and amplitude into account, which can be done using the mass conservation.

$$\rho_{in} v_{in} A_{in} - \rho_{out} v_{out} A_{out} = \frac{\partial (\rho V_0)}{\partial t} \qquad \text{eq. 2.28}$$

The momentary mass flow in and out of a volume determine the change in stored mass. With this definition, a process can be defined as quasi-steady if the change in stored mass is rather small. Cycle averaging the magnitude of the change in stored mass, and assuming an ideal gas, λ can be derived to equation eq. 2.29. For a more detailed explanation the reader may be refer to [17].

$$\lambda = \pi St = \frac{2\Delta p}{\gamma p_0} St \qquad \text{eq. 2.29}$$

For the calculation of the Strouhal number, a correction factor has to be introduced as the pressure pulses in the turbine are not sinusoidal, but have a very pronounced peak when the valves open. Therefore, the opening time of the valve is considered and, if unknown, approximated by $\Phi = 1/3 \, T_{tot}$. A flow with a λ close to unity will be highly unsteady and cannot be viewed as quasi-steady. All approaches presented so far are limited to a cycle average value and are not able to determine a timely resolved number. This can be achieved by the using a proposition of [8].

All of the numbers introduced above include a characteristic length representing the dimension of the turbine to enable the comparison of different sizes. If the focus is put on the nature of the pulse itself, two additional numbers are of interest. Firstly, the pressure gradient to describe the transient behavior versus time and, secondly, the specific gradient to describe the pressure pulse intensity in relation to the turbine speed (see equation eq. 2.30).

$$\Gamma(p) = \frac{dp}{dt} \frac{1}{n_{TC}} \qquad \text{eq. 2.30}$$

2.5 Virtual Engine Development

The complexity of engine development has increased continuously over the last decades as the engine has become an integral part of the vehicle. Constraints concerning cost, packaging, performance, efficiency, available fuels, emissions and consumer desires drive today's highly virtual development process, which rests on three main pillars.

1D-Simulations are used in the early phase of the development process to investigate different configurations like the charging concept (combination of multiple turbocharger, mechanical or electric chargers) or the exhaust gas after treatment system (catalyst light off, dimension of surface available for SCR) considering stationary or transient behavior. Another big field of interest is the prognosis of engine emissions during drive cycles like the Worldwide harmonized Light Vehicle Test Procedure (WLTP) or a cycle suitable to estimate the Real Driving Emissions (RDE). Generally, the fast calculation time enables the user to evaluate a large quantity of solutions or long transient processes while relying on a number of assumptions and simplifications. To account for these, a very good calibration of the models to accurate measurements is required in order to obtain a predictive system with a high quality of the results.

3D-CFD Simulations typically enable a detailed view into the engine processes: for example, the optimization of an injector spray pattern for a better mixture homogenization, modification of the intake channel and cylinder geometry for optimized charge motion or improved valve timing for an accurate scavenging process. With regard to the calculation time and the effort to build up new models, the areas of application and the number of configurations investigated are still limited and have to be chosen carefully. 3D-CFD simulations have become increasingly popular due to the direct integration into CAD software and the focus on user friendliness. As DNS is not an option for internal combustion engines, the models employed have to be suited and calibrated as well. However, the calibration effort is reduced compared to 1D-Simulation as a lot of the phenomena (residual gas in cylinder, mixture formation etc) are calculated and not modeled.

The role of the test bench has changed drastically with the popularity of simulation tools. Before, measurements where the only tool available for engine development and, therefore, optimization had to be done in hardware and was very costly and time-consuming. The focus has now been shifted strongly to perform fewer measurements but increase the accuracy to not only deliver calibration data for the simulations but also to develop new models and validate the simulation. It is, of course, evident that without accurate test bench data, the meaningfulness of simulations can be reduced remarkably.

3 Simulation Environments

3.1 QuickSim

QuickSim is a simulation tool dedicated to 3D-CFD of internal combustion engines, developed at the Research Institute of Automotive Engineering and Vehicle Engines Stuttgart (FKFS) and the Institute for Internal Combustion Engines and Automotive Engineering (IVK) by the University of Stuttgart. A detailed report on the applied methodology is given in [15], so the following introduction will be brief.

QuickSim follows a holistic design with the aim of aiding virtual engine development. In contrast to most commercially available codes, it is not limited to solving the fluid dynamic equations; instead, it is designed as a virtual test bench, generating data directly comparable to measurements. Optimization iterations of numerous parameters can therefore be run with an engine-specific target like better fuel consumption, efficiency or maximum power.

3.1.1 Methodology and Approach

Calculation time plays an important roll in the application of simulations in virtual engine development, especially in 3D-CFD, with discretization being one of the main influencing factors. By halving the length of the cell edge, the cell number increases by a factor of eight. A suitable maximum time step Δt for a given mesh size can be calculated with the help of the Courant number Co derived from the local convective velocity $\bar{v}$ and the cell discretization length l_D.

$$Co = \frac{\bar{v}\Delta t}{l_D} \leq 0.7 \qquad \text{eq. 3.1}$$

© Springer Fachmedien Wiesbaden GmbH, part of Springer Nature 2020
A. Kächele, *Turbocharger Integration into Multidimensional Engine Simulations to Enable Transient Load Cases*, Wissenschaftliche Reihe Fahrzeugtechnik Universität Stuttgart, https://doi.org/10.1007/978-3-658-28786-3_3

The distance traveled per time step of a fluid particle cannot be longer than the cell edge: in other words, the fluid particle cannot 'skip' a cell but has to be 'detected' at least once in every cell. A coarser cell discretization enables consequently the use of larger time steps.

The 3D-CFD simulation of internal combustion engines is especially complicated due to the characteristic turbulent length, with the integral length representing the largest vortex and the Kolmogorov length representing the smallest vortex. Both are time dependent and especially the Kolmogorov length is so small that a DNS requires a unfeasible number of cells (see 2.3.5). Instead, additional models are implemented which are specialized on the processes inside internal combustion engines and therefore allow for a relatively coarse mesh [15].

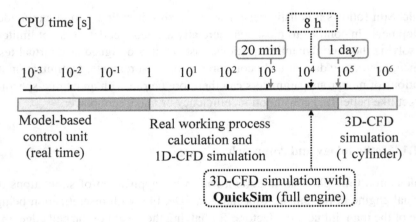

Figure 3.1: Time scale of different simulation approaches [60]

Figure 3.1 shows an exemplary time scale for the simulation of a full four-cylinder engine, with a total calculation time of 8 hours in QuickSim . Naturally, this value is dependent on the number of cylinders, the displacement, operating point specific variables like the amount of injected fuel and optional models like post-oxidation.

With a simulation time of 6-8 hours per engine working cycle, QuickSim is able to calculate multiple consecutive cycles for a stationary operating point.

This is crucial to eliminate any influence of the initialization and guarantees a 'converged' initial state at the beginning of each working cycle; otherwise the amount and location of cylinder residual gas or the storage of fuel in the intake system are unknown and have to be estimated from the test bench or 1D-Simulations. The differences between the first cycles can be substantial, as shown in [15].

In QuickSim, the extent of the simulated engine domain can range from the single cylinder, with only the intake and exhaust channels, to a full engine model with multiple cylinders, airbox and exhaust runners as depicted in Figure 3.2. The inclusion of a turbocharger will be presented at length in this work.

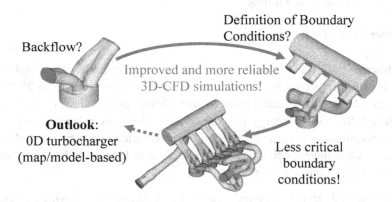

Figure 3.2: Simulation domain increase for reduced dependency of boundary conditions [60]

It was found in the past that it makes a significant difference for the gas exchange process, and the subsequent combustion, whether one cylinder is simulated or the cylinder interaction of the full engine is accounted for [61]. Therefore, the general approach is to extend the simulated domain as far as possible. In cases with very high demands concerning the calibration accuracy, this can lead to a full representation of the test bench [16, 47]. The size of the simulated domain not only determines the number of cells, but has a large impact on the simulation stability as boundary conditions with less time dependency and a uniform flow pattern (e.g. fully developed pipe flow without recirculation zones or reversed flow) are numerically superior.

QuickSim's characteristics can be summarized as follows:

- Engine-specific models enable coarse meshes with larger time steps
- Domain extension up to a multi-cylinder engines with complete air path
- Designed to deliver results directly comparable to the test bench
- A tool for the virtual engine development

3.1.2 Engine Specific Models

Cylinder Wall Heat Transfer

The description of heat transfer from the gas to the cylinder walls and vice versa is absolutely vital for an accurate simulation as it adds up to 25 % of the released heat at full load and more than 30 % at part load [27]. The losses therefore influence the energy balance of the entire system heavily and have to be determined as accurately as possible. A detailed overview of modeling approaches and their specific advantages is given in [15], so only a short summary will be given in this work.

The heat transfer zone is pictured as a thermal boundary layer, which consists of a turbulent main region and a laminar sublayer close to the wall. The actual calculation of the heat transfer is very difficult because it needs to take combustion engine specific phenomena into account (flame quenching, oil or fuel film etc). Furthermore, the existing models for 3D-CFD applications are commonly calibrated for simple wall geometries and flow structures. Inside the cylinder, the flow can be highly turbulent, reflected by the reduction of the laminar sub layer during combustion down to below 0.05 mm [15]. The consequently required mesh refinement is not feasible with current (2019) computational power.

In the real working process analysis, a lot of effort has been put into the development of phenomenological approaches, with a limited number of variables, specialized for the use in internal combustion engines. Woschni's correlation was introduced for diesel engines, but could also be used for gasoline engines in 1967 [64]. Hohenberg adapted the formulation in 1979, enabling a significant improvement for the gas exchange phase of a working cycle [29]. However, this formulation did not take the combustion into account and consequently underestimated the heat losses during the compression and expansion stroke. In the early 90's, Bargende proposed an additional combustion term that takes the flame propagation into account by dividing the cylinder volume into a burned and unburned zone [5]. Hohenberg's and Bargende's models have been validated and well used over the past decades, and are adapted in QuickSim to work on a discretized cell grid. As the cell grid is distorted during compression, an additional correction term has been implemented for these distorted cells.

Injection and Fuel Modeling

Injection of any kind (for example fuel, water or urea solution) is modeled by the initialization of fluid droplets inside the simulated domain. The calibration of the models employed as well as a detailed sensitivity study for the injection, has been carried out by Wentsch in [60]. A cone-shaped volume is defined for each nozzle of the injector in which droplets of different size and velocity are initialized. The diameter of each individual droplet is chosen according to the Rosine-Rammler-distribution with a control algorithm to obtain a desired Sauter-Mean-Diameter (SMD). The droplet velocity is determined as a function of the injection pressure and an empirical loss coefficient. The background gas flow and the liquid droplets are modeled as an Euler-Lagrangian two-phase flow and interact with each other. Several submodels like the droplet break-up according to Reitz-Diwakar or the splashing model according to Bai are employed. Droplets evaporate and change into the gas phase according to the pressure of saturation and the boiling curve. Liquids can be chosen as single or multi-component and are therefore able to replicate a complex fuel with a surrogate mixture of a comparable boiling curve.

Combustion and Working Fluid Properties

Combustion or oxidation of fuel is an imminent part of the internal combustion engine, requiring mechanisms to describe the thermodynamic fluid properties before, during and after this process, which can be difficult and time consuming. The most critical point is the chemical composition of the fuel and, in consequence of the burned gas. For single component fuels with only one element (e.g. hydrogen), 40 chemical reactions with 8 intermediate and final products have to be considered already. With increasing number of elements (e.g. methan or iso-octane), the number of chemical equations increases into the hundreds and thousands. For commercially available fuels, in which hundreds or even thousands of different species are blended, a numerical solution becomes unfeasible. To counteract this, reduced mechanisms, as shown in [58], have been developed.

However, even with these reduced mechanisms, the number of chemical equations is too high, leading to the approach introduced by Chiodi [15]. From a virtual single component fuel ($C_nH_mO_rN_q$) with a known lower heating value (LHV), the thermodynamic fluid properties of the burned gas are calculated under the assumption of eleven final products for a wide range of pressure, temperature, EGR concentration and lambda. This is done offline in the preprocessing and the information is stored in a database, from whence it can be retrieved quickly during simulation.

The heat release of the engine is divided into different submodels (e.g. flame propagation, diffusive combustion, volume reaction, post oxidation) which can be run parallel, enabling, for example, the evaluation of the post-oxidation contribution. The flame propagation model will be briefly described below (for further information see [15]), as it is the dominating process of the engines presented in this work.

The combustion is initialized close to the spark plug with the help of a spherical 'spark-kernel', which can be viewed as the plasma. The combustion propagates into the unburned surrounding gas with the laminar flame speed S_{lam} until the flame front has reached a dimension comparable to the turbulent vortices. After this point, the flame front propagates with the turbulent flame speed S_{turb} which is significantly higher than S_{lam} due to the turbulence-related wrinkling and stretching. The relation between laminar and turbulent flame speed is implemented using the wrinkling factor K, describing the increase of the flame front area due to the influence of turbulent eddies.

$$K = \frac{A_{f,lam}}{A_{f,turb}} = \frac{S_{turb}}{S_{lam}}$$
eq. 3.2

The flame propagation is modeled according to a local 2-zone approach, which divides each cell into a burned and unburned part (similarly to the real working process analysis) [66]. This is done with the help of a scalar progress variable ranging from zero (unburned) to one (fully burned). The turbulent flame speed can thereby be calculated with the properties (temperature, lambda, egr, humidity) of the unburned mixture within the cell.

3.2 STAR-CCM+

STAR-CCM+® is a commercially available simulation software distributed by Siemens. The user-friendly yet powerful environment offers integrated geometry analyses and repair functions as well as automated meshing. For setting up an individual simulation, the user can choose from a multitude of available models, all of them designed to work highly parallel and therefore well suited for large meshes.

In the present work, the software is used to calculate aerodynamic performance maps for a turbocharger turbine under stationary conditions. A script is run via the macro function in order to automate the repetitive tasks of adjusting the operating point. At a later point the simulations are run under transient conditions and serve as a reference case, against which the integrated 0D-Turbocharger will be validated. Detailed information on the used models for these simulations can be found in appendix A1.1

3.3 GT-Power

GT-Power is a module of GT-Suite distributed by Gamma Technologies that specializes in the 1D-Simulation of internal combustion engines. However, with additional input it also offers the option to simulate specific parts (e.g. intake channels) as 3D-CFD. It will be used in this work as a 1D reference for the virtual hot gas test bench and in the following application example. The employed models are listed in the Appendix A1.3.

4 Turbocharger Integration in QuickSim

4.1 State of the Art for Turbocharged Engines

It is common for 3D-CFD simulations of turbocharged internal combustion engines to limit the simulated domain on the intake side to parts of the intake channels and runners, sometimes including the airbox and throttle valve. The same principle is applied to the exhaust side where the boundary conditions are usually applied after the exhaust channels or the runners but always before the turbocharger (see Figure 4.1). In the figure shown, an external waste gate is integrated, requiring a second exhaust boundary condition.

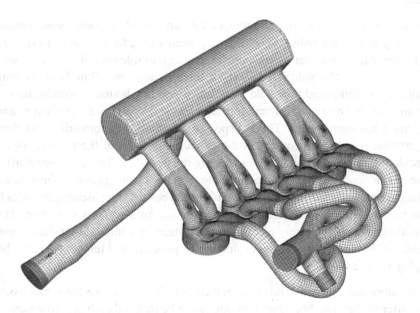

Figure 4.1: Four-cylinder engine with boundary conditions [60]

© Springer Fachmedien Wiesbaden GmbH, part of Springer Nature 2020
A. Kächele, *Turbocharger Integration into Multidimensional Engine Simulations to Enable Transient Load Cases*, Wissenschaftliche Reihe Fahrzeugtechnik Universität Stuttgart, https://doi.org/10.1007/978-3-658-28786-3_4

Pressure and temperature boundary conditions can be taken either from 1D-Simulations, comprising the entire system (including turbocharger, air filter, exhaust gas after treatment etc.) or from measurements on the test bench. In both cases, the availability of high quality boundary conditions is limited during the early development stage when no test bench data is available and consequently the 1D-Simulation can hardly be calibrated.

4.2 Benefits of an Integrated Turbocharger

Insufficient quality of boundary conditions is a major concern for the accuracy and reliability of 3D-CFD simulations as they strongly influence quality and stability.

On the test bench, pressure and temperature are usually a point measurement, resulting in a scalar value. Special arrangements like, for example, a pressure ring line, can help to get a averaged value over multiple spatial points but will still deliver a scalar value to describe a three-dimensional flow field. In consequence, the imposed flow field at the corresponding boundary condition will be uniform and one-dimensional, introducing the first cause of error before any equation has been solved. This is especially difficult for geometries with flow separation (e.g. sharp corners in the exhaust runners and the air box) or recirculating flow (strong pulsations causing the flow direction to change during one engine working cycle). In addition to this, the strong transient phenomena introduced by the valve openings require time-resolved measurements which can be very complex (e.g. temperature in exhaust runners). By definition, 1D-Simulations also offer only a scalar value instead of a three-dimensional flow field. An advantage is that the evaluation of pressure and temperature can be highly time-resolved.

If the flow domain is extended to include the air path before the compressor and after the turbine, boundary conditions can be reduced to the environmental. These are stationary, quasi one-dimensional and can be measured with high precision.

In some cases, the subject of research is not the cylinder itself but the phenomena occurring in the exhaust runners or after-treatment system. As these phenomena occur close to the turbine inlet, a fixed boundary condition before the turbine is likely to introduce an error. One example is a desired post-oxidation in combination with strong scavenging. In this case, the cylinder is operated with a combustion under rich conditions and strong scavenging of fresh air through the cylinder into the exhaust [26]. Adjacent cylinders will therefore alternately push rich exhaust gas and very lean fresh air into the exhaust where they are mixed and the remaining fuel is oxidized, increasing the enthalpy before the turbine and thereby the transient response.

The integration of a turbocharger allows for a full replication of the exhaust gas after-treatment system. The flow through the turbine is highly pulsed due to a minimized exhaust runner length and swirling as the turbine exiting flow angle is dependent on the mass flow and the speed of the turbocharger. Additionally, the sharp corners of the routing create a complex flow structure that can hardly be measured but requires intensive 3D-CFD simulations to optimize, for example, the urea injection into the SCR catalyst.

While variable turbine geometry (VTG) has become popular in diesel engines and in some high class gasoline engines, the bypassing of the turbocharger via a waste gate is still the standard for mass production gasoline cars. The valve can be located inside the turbocharger housing (internal waste gate) or in external parts coupled to the main exhaust system (external waste gate). Integrating the turbocharger and the waste gate enables the investigation of the flow and a possible negative impact on the turbine performance or NVH (Noise, Vibration, Harshness) behavior.

As the turbine flow is highly pulsed, the volume and shape of the turbine housing can influence engine performance significantly. Not only the design choice between single or twin scroll, but also the filling and emptying effects (transient storage) dependent on the engine and turbocharger operating point determine the performance. With an implementation of the turbine housing, the dependency can be fully captured.

The biggest advantage of the integrated turbocharger is the ability to capture the interaction between the turbocharger and combustion engine. Both machines influence their respective operating points and their mutual boundary conditions. The coupling enables a detailed energy balance of the entire system under steady and transient operating conditions. A change in the operating point of the internal combustion engine will consequently cause a change in the operating point of the turbocharger and vice versa. In consequence, the coupling is the key to transient load simulations over hundreds of working cycles. This can extend the use of 3D-CFD into a domain that was previously only achievable on the test bench and in 1D-Simulations. The high accuracy and detailed analysis can be used to understand complex phenomena like the turbolag or emission formation during transient load changes.

So far, the following benefits of the integrated turbocharger have been presented:

- Turbocharger and engine form a closed system requiring only environmental boundary conditions

- Improved simulation quality (e.g. more accurate simulation of scavenging through reduced influence of boundary conditions)

- Spatial and timely resolution of thermodynamic properties and species concentration in an extended simulation domain allowing further investigation like post-oxidation

- Detailed analysis of exhaust gas after treatment (e.g. swirl from the turbine influencing the urea injection and SCR catalyst performance)

- Complex effects of waste gate flow, including NVH analysis

- Realistic representation of storage and wave propagation effects in turbine housing

- Fully captured interaction between turbocharger and engine, enabling transient simulations over a large number of working cycles

4.3 Requirements for an Integrated Turbocharger Model

A potential model for an integrated turbocharger has to fulfill multiple requirements to work in an internal combustion engine environment as well as within
the established QuickSim framework and design purpose. The additional simulation time is critical, especially when targeting transient simulations of up to
multiple hundred working cycles. To ensure that the integrated turbocharger
can be used in virtual engine development and is not limited to purely scientific
purposes, simulation time has to remain in the order of magnitude to justify the
benefits given in 4.2.

As introduced in Chapter 2.4.2 the pulsating flow is characteristic for turbochargers in automobile applications. To determine the maximum range concerning amplitude, frequency, and the characteristic numbers given in 2.4.3, a
study of the exhaust pressure trace is carried out with multiple engines at different speeds. The sensors are located close to the flange of the turbine housing,
therefore interaction between the cylinders is fully captured. Pulsations are
more severe for the turbine than for the compressor, therefore the present work
will focus on those.

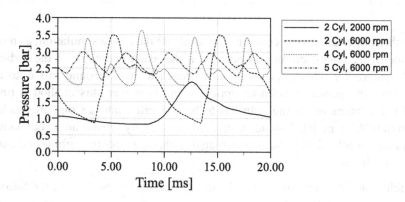

Figure 4.2: Example pressure traces close to turbocharger

Figure 4.2 shows the exhaust pressure traces of three different engines for Wide
Open Throttle (WOT) conditions. The examples have been selected, with the

target of a maximum pressure gradient. It can be observed, that the pulse frequency and amplitude are clearly connected to the number of cylinders and the engine speed. The pulse shape differs from case to case with the five-cylinder engine coming close to a sinusoidal function. In the case of the two-cylinder engine, the gradients of the rising and falling flank are significantly different and the four-cylinder engine even shows a secondary peak due to the resonance design of the exhaust system.

As the cases show, represent extreme operating conditions for a turbocharger, they are used to define the requirements under which the models have to operate stable and accurate. Among the traces displayed the four-cylinder shows the highest level of unsteadiness in combination with the turbocharger, making them the maximum requirements displayed in Table 4.1.

Table 4.1: Maximum requirements derived from figure 4.2

	Amplitude	Pressure Gradient	Sp. Pressure Gradient	Strouhal (Volute)	λ (Volute)
	bar	bar/s	bar/rev	-	-
Max.	2.5	2450	1.1	0.65	0.44

Another important aspect concerns the available data for a simulation, as most turbocharger manufacturers do not reveal the exact geometry of the turbine and compressor wheel. A common approach to work around this is to reconstruct the geometry from an optical scan. However, this method is time and labor intensive, as the scan has to be converted into a Computer Aided Design (CAD) model. Instead, manufacturers usually release an aerodynamic or thermo-mechanical performance map that characterizes the turbine and compressor wheels.

Concluding the requirements for the integration of a turbocharger, the following points have to be considered:

- Simulation time has to remain feasible

- Models have to be stable under engine typical pulsating conditions

- The models have to be designed to work with the data available

4.4 Approaches for an Integrated Turbocharger

Multiple approaches can be taken to implement a turbocharger, differing in simulation effort, accuracy and level of details [32].

1. The first approach uses an engine mesh similar to the one introduced in the Chapter 4.1. Instead of applying fixed boundary conditions at the manifolds, a one-dimensional model (of the turbocharger and the geometry not represented by the 3D mesh) calculates them for each time step. Therefore, the common boundary conditions (at the transition from 1D to 3D) will react to changes of the turbocharger and combustion engine. Additional simulation time is very low, as the computational effort of the 1D-Model is very low compared to a three dimensional multi-cylinder engine. On the downside, 3D-flow information before the compressor and after the turbine is lost, which can be of interest (e.g. for exhaust gas after treatment).

2. The second approach is to include the entire intake and exhaust manifold into the three dimensional domain, leading to a slightly increased computation time due to the higher number of cells in the mesh. The turbocharger is modeled via a map-based zero-dimensional approach. Due to the interruption of the mesh by the 0D-Turbocharger, special attention has to be paid to mass and energy conservation between the two boundary conditions. This includes a link to ensure that the species concentrations leaving the mesh are also reentering the mesh. Several derivatives of this approach can be designed, offering, for example, the choice whether the turbine volute should be part of the three- or the one-dimensional zone

3. In the third approach, the flow-related geometry (air path) of the turbocharger is represented in the 3D-zone but without an actual rotor. The correct energy balance is calculated via the turbine and compressor performance maps and implemented as numerical source terms saving the computational effort for a rotating mesh or a frozen rotor approach. A beneficial effect is the continuous modeling of the flow path without interruptions (like, for example, in approach 2). This automatically ensures mass conservation.

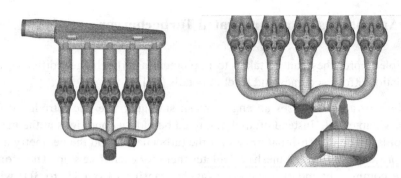

Figure 4.3: Approach 1 (left) and 2 (right) for the integrated turbocharger

4. The fourth and, by far most time-intensive approach, is the direct representation of the turbine and compressor inclusive the respective rotor in the
   three-dimensional domain. The time consumed by this simulation is significantly higher than for all of the previously described approaches, as the
   rotor speed is determined by the maximum physical time step. The allowed
   time step size for turbocharger simulations is closely linked to the speed
   and usually chosen between 1 and 5° of turbine shaft rotation.

Assuming 2° of rotation, this can lead to an increase in computation time
for the entire model of a factor of 10 and more. In the engine simulation,
a time step of 0.5° CA at 3000 rpm equals 2.78 *10-5 s per iteration while
the turbocharger requires a time step of 3.3*10-6 s at the corresponding
100,000 rpm. In order to save time, it is also possible to split the large simulation (including the cylinders, turbine and compressor) into several smaller
ones which are connected via virtual boundaries at the cost of many parallel
resources.

In addition to figure 4.3 and figure 4.4 a comparison between the approaches
can be found in tabular form in Appendix A1.4

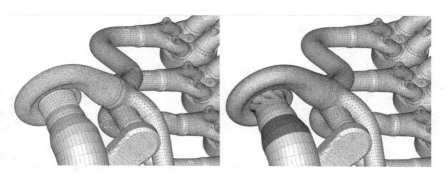

Figure 4.4: Approach 3 (left) and 4 (right) for the integrated turbocharger

Table 4.2 gives an indication about the calculation time for one working cycle of a two-cylinder engine on a single core processor (Intel Xeon E5-2643 v4) for the different approaches.

Table 4.2: Estimated calculation time (per cycle) for different approaches

3D-CFD Engine with 1D-System (Case 1)	5:00 h
3D-CFD Engine with 0D-TC (Case 2)	5:40 h
3D-CFD Engine with 3D-CFD TC (Case 4)	60 - 70 h (estimated)

One of the main goals of the present work is to enable transient engine simulations with more than 100 consecutive working cycles in a time aceptable for virtual engine development. Considering today's (as of 2019) computing power, this can only be achieved with the approaches 1 to 3. Approach 3 is believed to be very difficult in terms of stability and the presumable benefit of mass conservation can be ensured in approach 2 with little effort. Therefore, approach 2 is implemented and from now on referenced to as QuickSim 0D-Turbocharger.

5 The Chosen Approach: 0D-Turbocharger

5.1 General Modeling and Implementation

An overview of the integrated 0D-Turbocharger into the 3D-CFD Environment of QuickSim is given in Figure 5.1. Continuous lines represent three dimensionally meshed parts, whereas dashed lines represent 0D modeling. Only environmental boundary conditions are required, enabling the simulation of the low pressure intake side, including, e.g., the air filter. On the high pressure intake side, the charge air intercooler, throttle and air box are modeled as well as a possible high pressure EGR system.

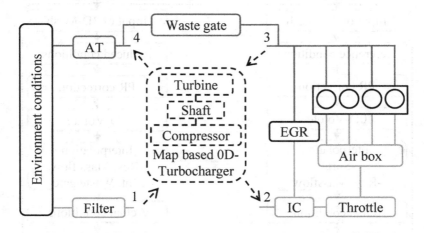

Figure 5.1: Concept QuickSim with 0D-Turbocharger

The exhaust path includes waste gate and exhaust gas after treatment with the ability to add additional technologies like water and urea injection or low pressure EGR. At the transition from 3D to the 0D-Turbocharger model (interfaces at positions 1 to 4), data is exchanged from the mesh to the 0D-Model or vice

© Springer Fachmedien Wiesbaden GmbH, part of Springer Nature 2020
A. Kächele, *Turbocharger Integration into Multidimensional Engine Simulations to Enable Transient Load Cases*, Wissenschaftliche Reihe Fahrzeugtechnik Universität Stuttgart, https://doi.org/10.1007/978-3-658-28786-3_5

versa. This information is processed and returned in the form of global information (e.g. as new shaft speed) or reimposed as internal boundary conditions.
The performance-map-based 0D-Turbocharger does not consider filling and
emptying phenomena of the included volume, hence these parts need to be
meshed or represented by additional 0D-Models. The concept can be extended for multiple turbochargers as well as any other kind of charging systems
like an e-charger or a turbo-compound system.

A detailed flow-chart including the programmed sub modules is given in Figure
5.2.

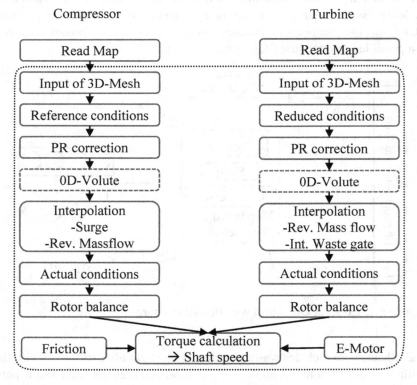

Figure 5.2: Flow chart of the integrated 0D-Turbocharger

All required performance map data is read and stored in the memory before the first iteration of the simulation to ensure a fast access. During the simulation (box with dotted line), flow data from the interfaces (pressure, temperature, species concentration, specific heat coefficients, etc.) is collected and converted to equivalent reference or reduced conditions with the help of eq. 2.22 and eq. 2.23 or eq. 2.24 and eq. 2.25.

Often a correction has to be carried out to ensure that the appropriate pressure ratio is used. The reason for this can be found in the divergent pipe after the turbine, which is used to convert kinetic energy into static pressure. When the performance map of the turbine is measured on the hot gas test bench, the pressure sensor is located after this diffuser (cross section area A_2 in the measurement plane) to reduce the disturbances from rotor blade passing. The static pressure measured is different from the one in the plane A_1 in Figure 5.3. In the simulation the user has the choice of including the diffuser into the mesh, adding a 0D-Model for the volume or ignoring the influence. Depending on the shape and size of the divergent section, all approaches can be valid. If the included volume is very small, its effects can be negligible, however, it is likely to be of interest for the simulation, especially in the case of an internal waste gate. In the present work, the divergent part is modeled in the mesh and the static pressure recovery from cross section A_1 to A_2 .

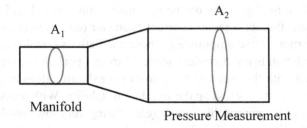

Figure 5.3: Schematic for calculation of pressure recovery in diffuser

The correction is carried out in an iterative procedure with the assumption of an isentropic change of state for an ideal gas and the conservation of mass and energy described by eq. 5.1.

$$p_2 = \sqrt[\frac{\gamma-1}{\gamma}]{\frac{bp_2^{-\frac{2}{\gamma}} - c}{a}} \qquad \text{eq. 5.1}$$

With the abbreviations:

$$a = -\frac{2c_p T_1}{p_1^{\frac{\gamma-1}{\gamma}}} \qquad \text{eq. 5.2}$$

$$b = \frac{A_1^2}{A_2^2} p_1^{\frac{2}{\gamma}} c_1^2 \qquad \text{eq. 5.3}$$

$$c = 2c_p T_1 + c_1^2 \qquad \text{eq. 5.4}$$

In the next module, the current shaft speed and pressure ratio of the compressor and turbine are used to determine mass flow and efficiency from the performance map with the help of an interpolation method. The extrapolation of the measured data is done in the pre-processing to ensure a fast and stable simulation. Nevertheless, sub-models have been implemented to handle surge or reversed flow and allow for calibration (see 5.2).

Due to the time discretization, the pressure ratio and mass flow may show slight oscillations depending on the operating conditions. This is not critical during a mass flow of a rather constant character (e.g. between two peaks at low engine rpm) as the amplitude of these oscillations remains small and the averaged solution is not time dependent. During a peak, however, the oscillations interact with the pressure wave from the exhaust valves magnifying the amplitude and putting stress on the simulation stability. With pressure ratios of the turbo machine approaching unity (e.g. during very low rpm in part load), the gradient of the mass flow increases. This can also lead to an amplification of the oscillation amplitude. To handle the cases in question, damping mechanisms are in place to reduce oscillations and ensure a stable operation.

In the case of a turbine, an internal waste gate is implemented. While all of the mass flow leaves the three-dimensional mesh and reenters after the turbine, the calculation of the available enthalpy and resulting rotor torque is adapted to account for the bypassed mass flow.

After the interpolation, the parameters derived are converted back to actual fluid conditions.

The rotor balance subroutine calculates the instantaneous power and torque values, as well es the resulting gas temperature of the turbine and compressor rotor outlet through the energy recovered from the expansion or required for the compression.

Torque balance of the shaft is calculated through the summation of the rotors and possible friction and/or additional torque, for example, from an electric motor.

5.2 Additional Models

5.2.1 Compressor Surge

The compressor surge line characterizes the maximum achievable pressure ratio in the performance map. Beyond this line, the theoretically required energy input into the flow is not possible, resulting in flow detachment and oscillating mass flow and pressure ratio in the compressor. Depending on the severeness of these pulsations, compressor surge can cause NVH problems, or, in extreme cases, destroy the machine. The phenomena is very complex, and therefore the description in the performance map is not fully sufficient (e.g. [19]).

Three different surge models have been investigated which are schematically displayed in Figure 5.4 with the surge beginning at the asterisk.

1. Constant mass flow

2. Linear decrease of mass flow

3. Exponential decrease of mass flow

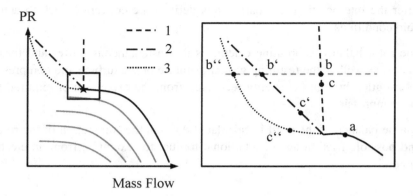

Figure 5.4: Implementation of surge model mass flow

For the constant case, mass flow is limited to the value at the maximum pressure ratio. This assumption is a very strong simplification and can only be justified very close to the surge line itself. To model an effect, similar to the partial back flow occurring physically within the rotor itself at deep surge conditions, the mass flow has to be decreased. Linear and exponential reduction of the mass flow through the compressor, thereby decreasing the pressure ratio and consequently moving the operating point closer to the surge line.

In the following thought experiment, depicted in Figure 5.4 on the right side, all three models react to the same operating conditions. Starting at point a, the pressure ratio of the compressor increases to a level b beyond the surge line. Depending on the model applied three different mass flows are calculated (b,b',b''). The exponential model results in the smallest value and the constant model in the largest one. In a third time step, the different mass flows lead to a difference in the pressure ratio between the models. The points (c,c',c'') differ also in their distance to the surge line. While the exponential model already operates very close to the original performance map, the linear and, especially, the constant model are still in significant surge.

In several applications, the exponential model has proven to be more stable and less sensitive to the physical time step size. Additionally, the transition at the surge line from the performance map to the model is smoother, because the gradient of the exponential model can be chosen, while the linear implementation introduces an inconstancy into the speed line.

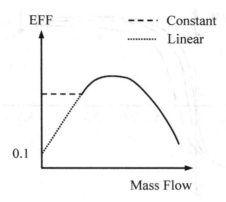

Figure 5.5: Implementation of surge model efficiency

The efficiency during surge is hard to measure and is usually only derived from 3D-CFD simulations. Nevertheless, two models have been investigated (see Figure 5.5), the first with constant efficiency and the second with a linear decrease to a chosen threshold. The influence of efficiency on simulation stability is remarkably low compared to the one of the pressure ratio. The assumption of a constant efficiency is physically not valid for the highly loss-driven flow occurring at surge, therefore the linear model is preferred. The threshold can be subject to calibration, however a default value of 0.3 has been found to work well on all cases tested.

5.2.2 Reversed Mass Flow

Performance maps are usually only provided for the first quadrant characterizing a flow motion from the turbine inlet towards the outlet. However, under certain engine operation conditions, the mass flow can change direction and the operating point is shifted into the third and fourth quadrant. In this case, a model has to be in place, that is able to approximate the behavior from the given first quadrant and a few additional calibration parameters. In other simulation environments, the extrapolation is done according to Figure 5.6.

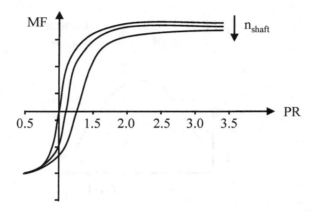

Figure 5.6: Turbine performance map with four quadrants adapted from [23]

This way of modeling can lead to very steep gradients of the mass flow for pressure ratios close to 1.0, especially for low shaft speeds. Therefore, very little change in the pressure ratio causes strong variation in the mass flow with the danger of creating an unstable simulation with a strongly oscillating flow. In a worst-case scenario the flow reverses the direction every iteration, making the results very sensitive to the time discretization.

To verify the extrapolation from Figure 5.6, an aerodynamic performance map of all four quadrants is calculated for a turbocharger with the help of 3D-CFD simulations (more details on the geometry used, simulation environment and modeling can be found in Chapter 6).

Analysis of the full aerodynamic performance map in Figure 5.7 did not show signs for a limitation of the negative mass flow in the range of the pressure ratios investigated; instead the mass flow seems to behave almost linearly for pressure ratios below one. As the reversed mass flow does have physical limitations due to the limited cross section area, it is believed to show this behavior for pressure ratios below 0.75.

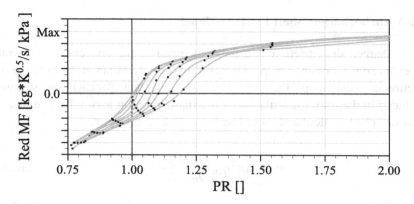

Figure 5.7: Turbine performance map of the turbocharger used in chapter 6

The influence of the shaft speed on the pressure ratio with zero mass flow can also be confirmed. At higher speed, this pressure ratio increases due to the losses induced through the higher circumferential speed. The high mass flow gradients for pressure ratios close to unity can be confirmed, as has already been shown in Figure 5.6.

Two different approaches were tested. While the first uses an extended performance map, generated during pre-processing, the second uses a conventional map and models the reversed mass flow during the simulation. The extended map is more user-friendly as it can be visualized. However, in the typical case where reversed mass flow data is not available, the extrapolation has to be performed without validation. In contrast to this, a modeled reversed mass flow of the turbine can be calibrated if the engine test bench data is available. For a good engine simulation model, exhaust pressure traces measured under operating conditions with reversed mass flow through the turbine can be used to calibrate the reversed flow modeling.

As both approaches have their justifications, the standard approach is to use the extended map with the user option to enable additional models designed to stabilize the area of very little mass flow and multipliers to scale the reversed mass flow quadrant.

5.2.3 Turbocharger Shaft Friction Model

Turbocharger shaft friction is usually included in the conventional thermo-mechanical performance maps. However, in the case of aerodynamic maps or friction-corrected maps, the friction has to be calculated for each time step and included in the torque calculation of the shaft. Friction loss in hydrodynamic bearings can be calculated according to the Petroff equation:

$$P_F = 6\pi^2 \eta_{oil} \frac{D_{sh}D_B^2 B_B}{D_B - D_{sh}} n_{TC}^2 \qquad \text{eq. 5.5}$$

With the P_F as power loss due to friction, η_{oil} the dynamic viscosity of the fluid, D_{sh} the outer shaft diameter, D_B the inner bearing diameter, B_B the bearing width and n_{TC} the speed of the turbocharger. As most of this data is unknown for the simulations, simplified approaches have been developed by several authors.

In the effort to determine the impact of various turbocharger design parameters, [54] promotes equation eq. 5.6 as a good estimation of the power loss due to shaft friction. It is supposed to be an average of different technologies (fully or semi-floating bearings), which is only dependent on the shaft diameter and the turbocharger speed.

$$P_F = 5.1E - 11D_{sh}n_{TC}^2 \qquad \text{eq. 5.6}$$

Instead of the shaft, [23] uses the rotor diameter to estimate the friction losses in the shaft.

$$P_F = 8.333E - 15(d_{wheel}[mm])n_{TC}^2[rpm] \qquad \text{eq. 5.7}$$

Several authors have introduced a second-degree polynomial with turbocharger specific constants. These have to be measured and are not generally available. For the two turbochargers investigated in [45] the constants are given in table 5.1

$$P_F = a_1 + a_2 n_{TC} + a_3 n_{TC}^2 \qquad \text{eq. 5.8}$$

Table 5.1: Friction constants for eq. 5.8 from [45]

	a_1	a_2	a_3
A	0	$3.98 * 10^{-4}$	$1.93 * 10^{-8}$
B	0	$5.66 * 10^{-5}$	$2.61 * 10^{-8}$

With the ongoing development of reducing the turbine shaft diameter, the ratio between wheel and shaft diameter is prone to change, increasing the need to constantly adapt the constant in eq. 5.7. As the shaft diameter is the main parameter of turbocharger geometry on the power losses due to shaft friction, the empirical approach in eq. 5.6 is implemented in QuickSim.

6 Validation by means of a Virtual Hot Gas Test Bench

6.1 Test Bench Setup and Simulation Environments

The modeling introduced in Chapter 5 will be compared to well-established 1D and 3D-Simulation tools and tested for accuracy, stability and computation time by means of a virtual hot gas test bench.

Hot gas test benches are commonly designed to measure performance maps or to calibrate heat transfer models for stationary operation. There are test benches designed to measure pulsating flow (e.g. [9]); however, it is quite difficult to obtain the required measurement accuracy. The validation of the modeling introduced in the present work is therefore performed through simulations. This ensures that the pressure and temperature pulses can be applied separately and allows for the creation of ideal pressure waves without the limitations and imperfections of a pulse generator.

In the case described here, the hot gas test bench is composed of the turbocharger turbine and straight inlet/outlet pipes of 500mm length each. The diameter of these pipes is maintained from the flanges of the turbine housing, as is standard for measuring thermo-mechanical performance maps. To restrict the investigations performed, to the turbine pulse analysis, a detailed aerodynamic performance map of the turbine investigated has been derived from the 3D-CFD model and will be used as input for the 1D-Simulation and QuickSim.

The commercially available software GT-Power (see Chapter 3.3) has been chosen, as it is widely used in the automotive industry and can therefore be used as a reference for 1D-Simulations. The models used are designed to handle pulsating flow and should consequently not limit the investigations in terms of stability.

© Springer Fachmedien Wiesbaden GmbH, part of Springer Nature 2020
A. Kächele, *Turbocharger Integration into Multidimensional Engine Simulations to Enable Transient Load Cases*, Wissenschaftliche Reihe Fahrzeugtechnik Universität Stuttgart, https://doi.org/10.1007/978-3-658-28786-3_6

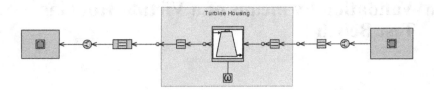

Figure 6.1: 1D-Simulation model of the virtual hot gas test bench

QuickSim is a tool dedicated to the virtual engine development, so it therefore requires a dummy engine with little simulation effort to run in the background (not shown). The virtual hot gas test bench, shown in Figure 6.2, is modeled with two pipes of the respective diameters and a small divergent section after the turbine (flow direction from left to right). Cell size is chosen to match the average exhaust system of a full engine simulation.

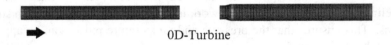

Figure 6.2: QuickSim model of the virtual hot gas test bench

In StarCCM+ (see Chapter 3.2), a full 3D-model of the turbocharger is implemented including the rotor, using the rotating reference frame option for the motion. Despite using the exact geometry, additional assumptions have to be made, e.g for the rotor tip clearance as a function of temperature and turbocharger speed. The influence of the cell discretization has been intensively investigated in stationary and transient simulations, to reduce the calculation time. The chosen mesh size of 275,000 cells is coarse compared to what other authors have used (e.g.[40]). This has been accepted, as the aerodynamic performance map, used in the 1D-Simulation and the integrated turbocharger, has been calculated with the same mesh. Within the quasi-steady regime, possible inaccuracies due to limited mesh refinement are incorporated in all simulation environments and do not influence the overall result.

Figure 6.3: 3D-CFD model of the virtual hot gas test bench

The 3D-CFD simulation is presumably the most accurate under transient conditions, as it includes most of the geometric details and is independent from additional input data (e.g. performance maps) or the quasi-steady assumption. It will serve as reference for the other two cases.

Turbine walls are modeled adiabatic during the generation of the performance map and during the transient simulations. The wall temperature in the inlet pipe is set to a fixed value of 1000K and to 900K for the outlet pipe. The turbocharger speed remains fixed for the duration of the pressure pulse.

The working fluid is modeled as a one-component flow with a modified nitrogen base model for the sake of computation time. Fluid properties like viscosity etc. are those of nitrogen, with the exception of the heat capacity imported from the QuickSim caloric file (see [16] for more details) as displayed in Figure 6.4. Herein, the assumption is made that the exhaust gas entering the turbine is derived from a complete combustion of regular gasoline (RON 98) and air under stoichiometric conditions.

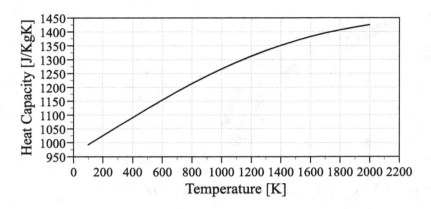

Figure 6.4: Heat capacity versus temperature at constant pressure (1bar)

A brief comparison of the three simulation environments is given in Table 6.1. Mesh size includes all fluid cells, in the case of QuickSim also the cells of the dummy engine. The simulation time has been calculated for a 50 Hz pulse on a single core of an Intel Xenon E5-2643 v4 processor. The temporal discretization remains constant for QuickSim, except for the stability investigation performed in Chapter 6.4.4. It is variable in the 3D-CFD simulation, depending on the pulse characteristic to reduce the cycle time while maintaining a stable simulation.

Table 6.1: Overview 1D, QuickSim and 3D-CFD simulation

	1D	QuickSim	3D-CFD
Mesh size / Cell number	-	76,000	275,000
Simulation time [min]	1	60	1,200
Temporal discretization [ms]	-	0.021	0.05 - 0.005

In all three simulation environments, the sensor locations have been defined according to Figure 6.5 where possible. In the 1D-Simulation, all sensors are reduced to a point; in QuickSim, all of the measurement locations are planar faces which are mass averaged. This is also true for the 3D-CFD case, except for the rotor inlet (position 3), which is cylindrical.

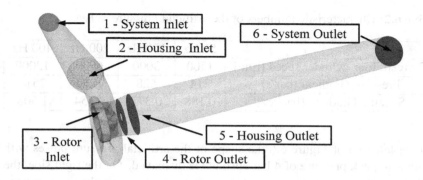

Figure 6.5: Positioning and nomenclature of the sensors

6.2 Artificial Pressure Traces

For a detailed analysis and validation of the models, substitute pressure traces
have been designed in order to represent the extreme values from the examples
in Chapter 4.2. The artificial pressure traces are sinusoidal for the sake of
scalability and are defined according to equation eq. 6.1.

$$p(t) = p_{start} + p_{amp}\left[1 + \sin\left(2\pi t f - \frac{\pi}{2}\right)\right] \qquad \text{eq. 6.1}$$

Table 6.2 provides an overview of the transient characteristic numbers for multiple frequencies at a 2 bar amplitude. For the reader's reference the corresponding speed of an ideal four cylinder engine is given as well. To reach the
extreme pressure gradients of the example traces, frequencies up to 400 Hz are
necessary. The chosen range from 50 Hz to 400 Hz is significantly broader
than what others consider (e.g. [2]) and extends the applicable range from
series to sports and racing engines. The temperature is kept constant throughout the pressure pulse to exclude the temperature influence on the the caloric
and sonic properties of the fluid. Other authors have found the temperature
influence to be significantly lower on turbine pulse behavior than the pressure
[2].

Table 6.2: Characteristic numbers of the artificial pressure pulses

	50 Hz	100 Hz	200 Hz	400 Hz
Corresp. four-cylinder speed [rpm]	1500	3000	6000	12000
Pressure Gradient [bar/s]	314	628	1257	2510
Specific Gradient [bar/rev]	0.188	0.377	0.754	1.508

On the left side of Figure 6.6 the shape of the artificial pressure pulses with a constant peak pressure of 4 bar absolute is displayed. On the right side, the 400 Hz pressure pulse is compared to the five-cylinder example given earlier.

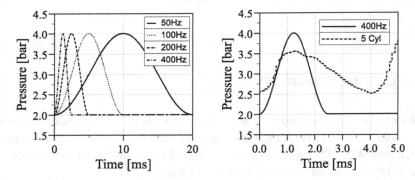

Figure 6.6: Artificial pressure pulses with five-cylinder example

6.3 Stationary Flow on the Virtual Hot Gas Test Bench

As the investigations at the VHGTB aim to compare the turbine performance only, any influence of the compressor and the inlet and outlet ducting of the turbine needs to be excluded from the analysis. This is done by a calibration process, ensuring identical turbine operating conditions for all three simulations.

In the calibration process, static pressure and the temperature at the test bench inlet are adjusted to achieve a pressure ratio of 2 bar and a housing inlet temperature of 1000K. The required adjustment is relatively small and always below 1 % but shall be addressed here nonetheless. By doing so, all turbines operate under identical conditions and, if using an aerodynamic performance map, should deliver identical results.

The operating conditions of the stationary flow have been chosen at approximately two bar static intake pressure and a fixed turbocharger speed of 100,000 rpm. The static pressure after the VHGTB is set to 1 bar absolute, allowing the turbocharger to operate in the middle part of the performance map, where the data fitting and interpolation work best.

The results are shown in Table 6.3, demonstrating that the input values for the turbine (PR, enthalpy and enthalpy flow) have been calibrated to similar values. The resulting mass flow and efficiency show an acceptable agreement for all simulation environments and the three pressure ratios investigated.

Table 6.3: Stationary calibration results of the VHGTB

	PR	Efficiency	MF	Enthalpy	Enthalpy Flow
	-	%	kg/s	kJ/kg	kW
1D-Simulation	2.00	60.5	0.146	195.7	28.5
QuickSim	2.00	60.0	0.147	195.5	28.7
3D-CFD	2.00	59.5	0.145	195.7	28.4
1D-Simulation	3.00	50.3	0.229	295.8	67.7
QuickSim	3.00	49.7	0.230	295.7	68.3
3D-CFD	3.00	50.9	0.241	295.8	71.4
1D-Simulation	4.00	44.0	0.314	360.9	113.2
QuickSim	4.00	43.5	0.314	360.8	113.5
3D-CFD	4.00	43.5	0.327	361.3	118.0

The remaining differences in mass flow are caused by the fitting algorithm during the generation of the aerodynamic performance map in the pre-processing. This is reflected in the fact that both, the 1D-Simulation and QuickSim, show the exact same behavior.

The differences in efficiency are believed to be caused by the interpolation from the discretized performance map during the simulation (e.g. linear, cubic, spline interpolation). No trend can be derived from the data gathered other than that the deviation decreases with higher pressure ratios. This is very likely due to the reduced gradient of the maximum efficiency line for higher pressure ratios.

6.4 Instationary Flow on the Virtual Hot Gas Test Bench

After the validation with stationary operating conditions, the VHGTB is run with transient pressure profiles, defined in accordance with Chapter 6.2. For the sake of readability, only the two most extreme cases of 50 Hz and 400 Hz pulse frequency are displayed in this chapter, as it has been found that the phenomena occurring are less pronounced for the 100 Hz and 200 Hz cases which have also been simulated.

The aims of these investigations are the following:

- Compare the system reaction for different simulation environments and validate the 0D-Turbocharger against 3D-CFD and 1D-Simulations which can be seen as the industry standard. This is achieved through a study on the influence of pressure wave frequency, amplitude and turbocharger speed

- Investigate different volute representations and find best practice

- Characterize and optimize stability of the 0D-Turbocharger in QuickSim

- Find appropriate time discretization values for all three simulation environments

Aiming to analyze and compare the 0D-Turbocharger, the turbine representation itself is the subject of interest. Consequently, it is critical to ensure identical conditions at the housing inlet and outlet in all three simulation environments. This rises the question of whether the inlet and outlet pipe should be included in the simulation at all.

By setting the system inlet boundary conditions very close to the turbine itself (no or a very short inlet pipe), the conditions at the housing inlet are fixed. At the same time, the flow is not able to develop a typical pipe flow pattern. It has also been found that, for high pressure gradients the simulation stability deteriorates as the solver control has difficulties imposing the boundary conditions onto a possibly contradictory flow field.

A longer inlet pipe, on the other hand, allows the flow to develop a typical pattern, but is prone to oscillations, especially for lower pulse frequencies where the pulse has enough time to travel multiple times through the system. Consequently, the desired pipe length is dependent on the frequency spectrum investigated. This is also reflected in the literature, where the chosen inlet pipe length varies between publications.

6.4.1 Influence of Pulse Frequency

System Inlet

Figures 6.7 and 6.8 show the pressure and mass flow traces at the system inlet for the extreme cases of 50 Hz and 400 Hz. Starting from the static calibration to an identical pressure ratio, all three simulation environments are subject to a pressure wave with a peak-to-peak amplitude of 2 bar.

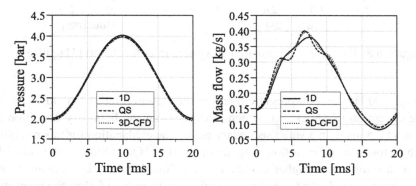

Figure 6.7: Pressure and mass flow at system inlet for the 50 Hz pulse

In the 50 Hz case, all three simulation environments follow the imposed pressure boundary conditions well; they differ solely in the calibration offset required to reach the desired pressure ratio at the turbine. Mass flow, calculated from the imposed boundary conditions, is different for all three simulation environments. The 1D-Simulation has a lower peak while QuickSim and the 3D-CFD show very good agreement. At the same time, the mass flow is oscillating more strongly in QuickSim. This oscillation has been found to be directly related to the intake pipe length and the pressure wave travel time. By superposition of multiple traces of different pipe length, they can be reduced to a point where they are no longer noticeable. The reason for these oscillations can be found in the control mechanism implemented in QuickSim to ensure that the simulation follows the boundary conditions. It has to be emphasized that this is an effect of the VHGTB and does not occur in the application with a full engine simulation.

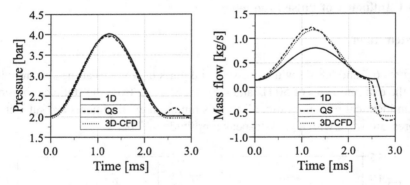

Figure 6.8: Pressure and mass flow at system inlet for the 400 Hz pulse

In the 400 Hz case, all pressure traces at the system inlet match closely, with the exception of QuickSim, which deviates from the boundary condition at around 2.5 ms with a small secondary peak. The explanation for this behavior can be found in the mass flow diagram on the right. QuickSim and the 3D-CFD show a significantly higher maximum mass flow than the 1D-Simulation. At the end of the pulse, all simulations obviously struggle to follow the boundary conditions and drop, rapidly reversing the mass flow dramatically. While the 1D-Simulation and the 3D-CFD allow for this rapid drop, it is intentionally forbidden in QuickSim as such a behaviour is prone to instability. In QuickSim the boundary condition is allowed to deviate from the input trace to ensure stability. As this effect occurs at the very end of the pulse, there is no influence on the pulse itself.

It can be concluded that, despite the stationary calibration effort, the simulation environments calculate a different mass flow at the system inlet for all pulse frequencies investigated.

Housing Inlet

As previously explained, the aim is to compare the turbine behaviour itself; consequently, all three simulation environments should show similar values at the housing inlet. In the 50 Hz case, presented in Figure 6.9, the differences are acceptable. The pressure is quite similar among the simulation environments, with QuickSim showing the explained oscillations. In the mass flow diagram QuickSim, and the 3D-CFD show a flat peak, where the 1D-Simulation shows a very pointed trace. A small shift is also visible when the 1D-Simulation and QuickSim are compared to the 3D-CFD.

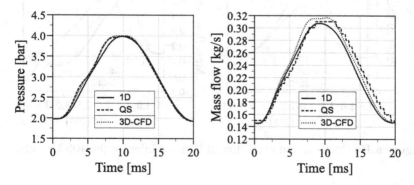

Figure 6.9: Pressure and mass flow at housing inlet for the 50 Hz pulse

In the 400 Hz case, shown in Figure 6.10, stronger differences become apparent. The 1D-Simulation does not detect the same pressure gradient at the beginning of the pulse as the other two, leading to a significantly reduced pressure peak. This is believed to be a consequence of the significantly reduced massflow at the system inlet. QuickSim and the 3D-CFD react quite similarly to the 400 Hz pulse with minor difference concerning the pressure gradient in the rising flank and the pressure maximum. Interestingly, the location of the pressure peak is similar for all three of them with a small delay for the 1D-Simulation.

The mass flow at the turbine housing, however, shows large differences between the simulations. The mass flow peak of the 3D-CFD is significantly higher than the other two. Additionally, the location of the mass flow peak varies remarkably. The most likely explanation for this behavior is the map extrapolation mechanism. The performance maps are often limited to a pressure ratio of 3.5. In this case, the range has intentionally been enlarged up to a ratio of 4.0, to deal with the high pressures occurring in the volute. However, the investigations show that a pressure ratio of 7 or 8 would be required in order to fully capture the behavior without extrapolation.

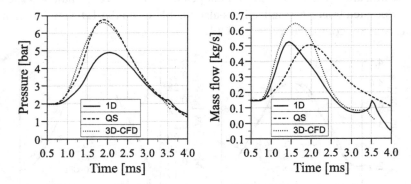

Figure 6.10: Pressure and mass flow at housing inlet for the 400 Hz pulse

The figures above show the difficulties of setting up identical conditions for all three environments, especially under high frequency pulses. In the 50 Hz case, QuickSim is closer to the reference case (the 3D-CFD), than the 1D-Simulation, which can be seen as the industry standard when calculating engine-turbocharger interaction. Despite the fact that the mass flow for 400 Hz is not identical, it will be accepted as an extreme case and further investigated.

Total Pressure at the Rotor Inlet

Turbine inlet area varies in size and shape for the different simulation environments. While it is a circular area for the end of a pipe in the 1D-Simulation and QuickSim, it is cylindrical in the 3D-CFD. Consequently, the pressure can no longer be compared at this measurement location. In a first approximation, the result of the different areas can be viewed as a nozzle, increasing velocity and decreasing pressure. To compare the simulations at the turbine inlet, consequently, the total pressure has to be evaluated instead of the static one. The results for the 50 Hz and 400 Hz pulse are shown in Figure 6.11.

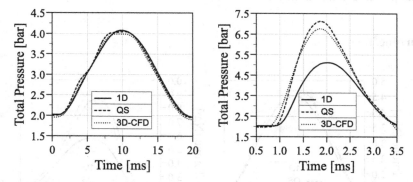

Figure 6.11: Total pressure at turbine inlet for 50 Hz (left) and 400 Hz (right)

In the 50 Hz case, only small differences can be observed, for example, the oscillation of QuickSim. The 1D-Simulation does show a more pronounced/sharp peak compared to the other two. The 3D-CFD is approximately 3 % lower at the peak, which can be attributed to the losses in the turbine housing. While they are calculated through fluid dynamics in the 3D-CFD, they are included in the performance map for the 1D-Simulation and QuickSim (the turbine housing wall is modeled without friction in this case). In consequence, the losses will be taken into account after the measurement location 'turbine inlet' for the latter two. The level of 3 % for the losses appears to be a reasonable value for this.

This picture changes drastically for the 400 Hz case, when the first signs of the pulse can be detected at the turbine (around 1.0 ms). The 3D-CFD detects an early increase in pressure, with a low gradient. This can be explained by the fact that the turbine inlet (cylindrical surface fed by a spiral) does not have a homogeneous total pressure field; instead the parts closer to the housing inlet will see the pressure increase before the part furthest away. The 1D-Simulation and QuickSim detect a very steep increase without the smooth transition, because the volute is modeled as a pipe with a discrete length. The total pressure gradient of QuickSim and the 3D-CFD increases similarly and results in a 3 % higher peak total pressure for QuickSim, just like in the 50 Hz case. The 1D-Simulation, however, shows the remarkably low total pressure gradient and total pressure peak already discussed.

Turbine Torque

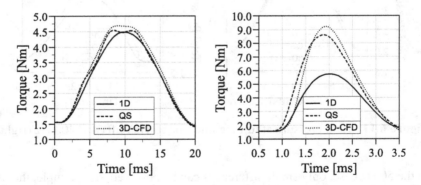

Figure 6.12: Torque at turbine inlet for 50 Hz (left) and 400 Hz (right)

The torque traces for 50 Hz and 400 Hz displayed in Figure 6.12 show the same trend as the stationary results in the previous chapter. The 3D-CFD calculates a slightly higher torque than the other two simulations. QuickSim and the 1D-Simulation, which share the same aerodynamic performance map, deliver quite similar results for the 50 Hz case indicating that the differences can be found in the generation of the performance map. In the 400 Hz case in the 1D-Simulation, the torque level is significantly reduced as were the pressure at housing inlet and the total pressure at the turbine inlet. As with the 50 Hz case, the aerodynamic performance map seems to underestimate the maximum torque. The reason can be partially found in the different mass flows through the turbine, also shown in Figure 6.10.

The results can be summarized as follows:

- For low pulse frequencies, the agreement is good, especially when considering that the oscillation in QuickSim is only an effect of the VHGTB and does not occur in an engine simulation

- No way has been found to ensure identical conditions at the housing inlet for the 400 Hz pulse, which is attributed to the individual controls of the boundary conditions

- High pulse frequencies highlight the differences between the models, causing stronger deviations.

- Between the two commonly used tools to simulate turbochargers, namely the 1D-Simulation and the 3D-CFD, remarkable differences occur at high pressure frequencies. QuickSim, which is essentially a hybrid between the two approaches, does perform significantly better for higher frequencies than the 1D-Simulation

6.4.2 Influence of Pulse Amplitude

Up to this point, the pulse amplitude has been constant at 2 bar. To evaluate the influence of the amplitude on the turbine response and the simulation stability, investigations have been repeated with an amplitude of 1 bar and 3 bar. Torque traces for all three amplitudes are displayed in Figure 6.13 for the 50 Hz pulse.

The torque evaluated here is strongly influenced by the pulse amplitude, as the enthalpy available for expansion increases. The trends identified at 2 bar pressure amplitude and a frequency of 50 Hz can be confirmed for 1 bar. 3D-CFD shows the higher torque values than QuickSim and the 1D-Simulation.

QuickSim also oscillates more strongly than the other two. For the 3 bar amplitude case, a double peak is visible in the 3D-CFD and in QuickSim, whereas the 1D-Simulation does not predict this behavior. It becomes obvious that the 1D-Simulation always computes a very smooth torque trace in the shape of an inverse parabola. The torque peak observed is significantly narrower than what 3D-CFD and QuickSim predict. Despite detecting a double peak trace, QuickSim and the 3D-CFD differ in their maximum torque. While the 3D-CFD calculates the first peak to be the maximum, QuickSim detects a higher value at the second peak. The cause for this behavior is believed to be found in the volute representation.

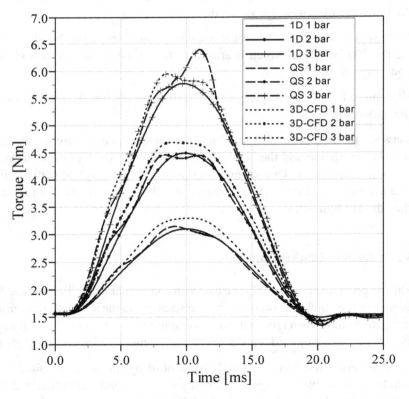

Figure 6.13: Torque for 1, 2 and 3 bar amplitude and a 50 Hz pulse

The 400 Hz pulse, displayed in Figure 6.14, does not show a clear separation between the groups of lines for each pressure amplitude. This behavior is mostly driven by the underestimation of the peak pressure for the 1D-Simulation already explained, which gets worse for higher pulse amplitudes. QuickSim and the 3D-CFD show a reasonably close behavior for 1 bar and 2 bar pulse amplitude. However, at 3 bar, the differences are remarkable. While the 3D-CFD has continuously shown the highest torque peaks, the behavior is not true anymore. QuickSim also detects the torque increase earlier than the 3D-CFD.

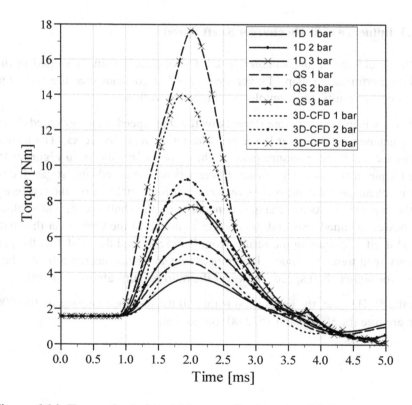

Figure 6.14: Torque for 1, 2 and 3 bar amplitude and a 400 Hz pulse

The result is that the differences between the simulation environments increase with higher pulse frequency and higher pulse amplitude. Both parameters contribute to a higher pressure gradient, which is a good indication for the stress that is put on the system. As shown in Figure 4.2 and Table 4.1 the worst case scenario for the pressure gradient is 2500 bar/s, corresponding to 2bar amplitude at 400 Hz. As has been shown, QuickSim can handle this with good results compared to the 1D-Simulation. At the same time, it becomes obvious, that with an amplitude of 3 bar (resulting in a pressure gradient of 3750 bar/s), the limits of QuickSim or the utilized quasi-steady assumption are reached.

6.4.3 Influence of Turbocharger Shaft Speed

With the change in turbocharger speed, the operating point is shifted in the turbine performance map. Lower shaft speed at constant pressure ratio and temperature results, for example, in a higher mass flow.

Efficiency is influenced more strongly by the shaft speed, as lower speeds have the efficiency maximum at lower pressure ratios, and vice versa. The torque traces for the 50 Hz pressure pulse at three speeds are shown in Figure 6.15. The torque traces show significant differences for the speeds investigated, despite using an identical pulse. Two reasons contribute to this: first, the efficiency of the expansion process changes depending on the shaft speed, and so does the power obtained. Second, this power is linked to the torque via the shaft speed itself. Comparing the results from QuickSim and the 3D-CFD, the previously seen trend continues. Differences in peak torque decrease in absolute terms for higher shaft speeds, however the ratio remains almost constant.

For the 50 Hz case, the conclusion is that all the previous findings at 100,000 rpm are true for 50,000 and 150,000 rpm as well.

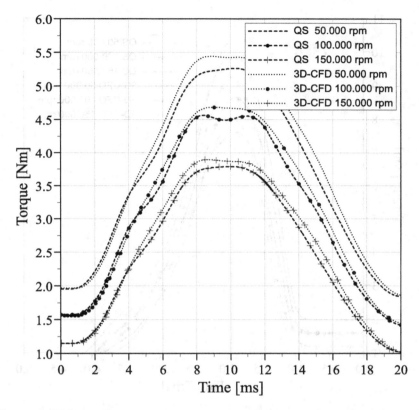

Figure 6.15: Torque trace for 50k, 100k and 150k rpm and a 50 Hz pulse

The similar figure of the turbine torque for a 400 Hz pulse can be seen in Figure 6.16, where the small delay between QuickSim and the 3D-CFD remains constant for all three shaft speeds. The higher maxima of the 3D-CFD appears constant for the different speed as well. In conjunction with the previously demonstrated change in amplitude, a large area in the performance map has been covered, indicating the validity of the quasi steady assumptions.

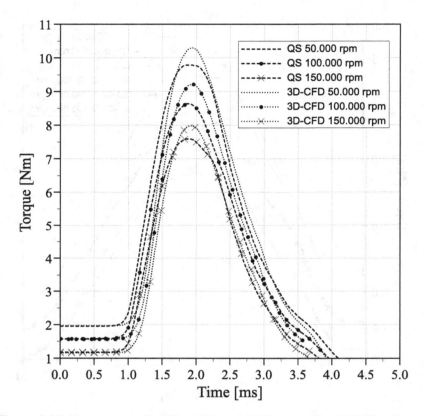

Figure 6.16: Torque trace for 50k, 100k and 150k rpm and a 400 Hz pulse

6.4.4 Simulation Stability

Numerical solvers are designed to find the solution for a system of equations and the respective variables. This concept can be applied to stationary problems, but also to transient, meaning time-dependent, ones. In the case of the 0D-Turbocharger, in conjunction with a combustion engine, there are limits to whether the solver is able to solve the system of equations for each consecutive time step. This does limit the stability of the simulation. Multiple factors contribute to the stability of a simulation; the most critical will be discussed with special reference to the 0D-Turbocharger.

Mass flow through the turbine is calculated by the 0D-Turbocharger model and enforced on the mesh face representing the inlet and outlet of the turbine. Depending on the mesh shape, fluid properties etc., a pressure is calculated for each cell by the solver. As mass flow and pressure of these cells are connected not only by the equations of gas dynamics, but also the turbine performance map, the system's ability to dampen possible excitation and limit resulting oscillations must be ensured.

QuickSim features a special type of manifold cell close to the boundary face to increase stability for fast changes. Mass flow or pressure is enforced on the boundary faces and the deviation from the target evaluated in the adjacent manifold cells. Usually, the manifold cell number and size do not influence the results, as the boundary is located far away from the engine. In the case of a 0D-Turbocharger, however, this can be critical, because multi turbine setups can feature especially short piping between the turbines. To determine the impact of the manifold cell size on the stability and accuracy of the simulation, the volume is varied.

To stress the simulation, the previously discussed 400 Hz pressure pulse with 2 bar amplitude is used. The sensor location at the turbine outlet is chosen as an indicator for the simulation stability, because the flow has to be reintroduced into the 3D-Mesh, which is prone to instability. The accuracy will be determined by how well the 3D-Flow field is able to follow the target values from 0D-Turbocharger model.

Figure 6.17 shows the mass flow at the turbine outlet for various cell sizes. The differences among them are within the width of the plotted line, leading to the conclusion that a substantially reduced or increased manifold cell size does not have a negative impact on the ability to follow the imposed boundary condition.

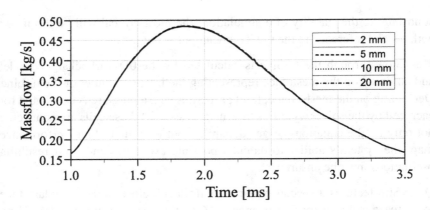

Figure 6.17: Influence of manifold length on mass flow trace

The physical time step chosen influences the computation time strongly and is often limited by the courant criteria (see eq. 3.1), describing the relation between cell size and physical time step. For a given grid and flow problem, a larger physical time step is typically more prone to instability, as the changes in flow structure between the iterations are larger. A series of simulations has been set up to determine the influence of the physical time step size for a cell size typically used in QuickSim engine simulations.

It has been found that, for a maximum pressure gradient of 2500 bar/s, corresponding to a pressure pulse of 400 Hz and 2 bar amplitude, (see Table 6.2) an physical time step per iteration of 0.02 ms is sufficient to guarantee stability. This can be viewed as a lower border for most pressure pulses relevant under engine operating conditions (as discussed in Chapter 4.2).

However, the temporal discretization does not only affect the stability of a simulation, but also the achievable accuracy, similar to the effects of spatial discretization. Figure 6.18 depicts the pressure at the turbine housing inlet for different physical time steps showing a delayed increase and a reduced pressure maximum for smaller time steps. When the time step is further decreased, the curves merge into one.

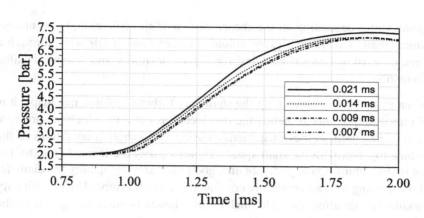

Figure 6.18: Influence of temporal discretization on a 400 Hz pressure pulse

For the use of the implemented models in an engine environment it has to be emphasized that the shown extreme operating conditions occur only at very few engines. The high pressure gradients are also only apparent during a short period of each working cycle. Nevertheless, the physical time step has to be chosen carefully to find a good compromise between stability and computational cost.

6.4.5 Comparison of Turbine Housing Models

For the volute, also referred to as turbine housing, literature agrees that the quasi-steady assumption of the turbine is not valid (e.g. [2, 3, 7, 11]). To account for this, two approaches are possible. Firstly, the turbine housing can be modeled via an additional one dimensional pipe or volume as an addition to the 0D-Turbine. The second approach is to include the volute in the three dimensional mesh, enabling the user to choose between different possible volute implementations. These can range from idealized geometries like simple pipes up to the detailed geometric models with guide vanes etc.

Previous chapters have shown that the 1D-Simulation has trouble capturing the pulse behavior for high frequencies. This might also apply to a one dimensional pipe modeling. The additional computation time required for the second option varies depending on the chosen degree of simplification, but is always rather small. Consequently, the volute is represented in this work through a three dimensional mesh, allowing for higher flexibility and a more detailed simulation.

When measuring or simulating the stationary turbine performance maps, it is of utmost importance to define the two locations between which the pressure ratio is measured, as they characterize the current operating conditions of the turbine together with the shaft speed. Typically, they are located close to, but not in the turbine housing. To obtain a good measurement quality the diameter of the housing inlet and outlet are continued as a straight pipe. Under stationary conditions, the effect of additional friction losses is small compared to the gains in measurement quality.

When applying the performance maps obtained to a simulation, care must be taken that the measurement locations of the pressure ratio are defined comparably, meaning at a location with an identical cross section area. As shown in Figure 6.19, the pressure is either measured at the end of the mesh (incoming or leaving flow) or at a virtual sensor somewhere inside the simulation domain.

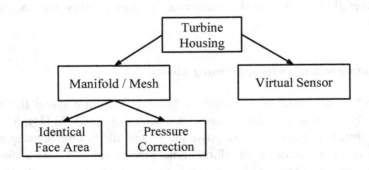

Figure 6.19: PR sensor concepts for the turbine housing

A virtual sensor offers the opportunity to place the pressure sensor at the location where it was positioned during the generation of the performance map. When obtaining the pressure from the last cells of the mesh, the respective face has to be of the same area as the cross section of the measurement. If this is not the case, a correction is required, for example via the mass balance and the conservation of energy with the assumption of an isentropic expansion.

Under transient conditions, the timing of the measured pressure trace becomes relevant. In this case, the virtual sensor (positioned before the turbine housing) is a disadvantage because it can introduce a faulty signal due to the difference in pulse travel length. For this reason, the concept of a virtual sensor is not further pursued in this work.

In some 1D-Simulations, the turbine housing is not considered to be a simple volume, but it is modeled as a pipe to maintain a characteristic wave travel length [2]. The pipe dimensions can be determined according to [11], with the pipe length l_{vol} and the pipe diameter d_{vol} determined using equation eq. 6.2 and eq. 6.3.

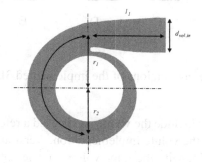

Figure 6.20: Definition of the volute dimensions in accordance with [11]

$$l_{vol} = l_1 + \frac{r_1 + r_2}{2} * \pi \qquad \text{eq. 6.2}$$

$$d_{vol} = \sqrt{\frac{4 * V_{vol}}{\pi * l_{vol}}} \qquad \text{eq. 6.3}$$

If the volute volume is unknown, an approximation for the pipe diameter is recommended:

$$d_{vol} = 0.9 * d_{vol,in}$$ eq. 6.4

If the described concept is applied to a three dimensional turbine housing mesh, it will result in a abrupt diameter change from the inlet pipe to the volute. While this might be acceptable in 1D-Simulations, as the induced losses can be deactivated, it is certainly not desired in 3D-CFD, because of the induced disruption of the pipe flow with vertices influence the main flow. Figure 6.21 shows different volute representations, which will be compared with respect to simulation stability and accuracy in comparison with the 3D-CFD of the turbine.

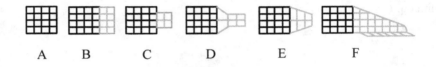

Figure 6.21: Different versions of the implemented 3D turbine housing

Version A does not include the volute and is used a reference case to determine the importance of the volute implementation. Version B is set up as an extension of the inlet pipe with the same volume as the volute, B1 with half of the volute volume. Version C is the implementation as proposed by [2] and [11] for 1D-Simulations. Version D is similar to version C, with the difference of a tapered section, allowing for a smoother transition. Version E is a constant taper, determined by the volute length and its volume. Case F is designed to be a closer approximation of the volute nature by removing the distinct volute length and replacing it with an outlet that enables the flow to leave the system over its entire length.

For stationary flow, the introduced implementations of the turbine housing differ in pressure ratio due to the variant pressure losses as expected. The sharp reduction in C, for example, causes higher losses compared to other solutions. As explained in 6.3, the previous models of the HGTB had to be calibrated to a stationary pressure ratio of 2.0; this procedure is repeated for all turbine housing models.

Figure 6.22 shows the torque traces for a 400 Hz pulse with 2 bar amplitude in comparison to the 3D-CFD simulation serving as a reference. This pulse has been chosen again for demonstration purposes as it is the worst case with the largest deviations. Under these conditions, the length of the volute is of remarkable significance to the timing of the torque trace. Versions A, B1, B, C show a delay in the order of their volute length. Version C, which has the volute length calculated in accordance with [11], shows the first signs of torque increase contemporaneously with the 3D-CFD at around 1.0 ms. At the same time, the height and timing of the maximum pressure peak are predicted more accurately than in versions A, B1 and B. Hence, the conclusion can be drawn that the calculated volute length is superior under the tested conditions when compared to an extruded mesh.

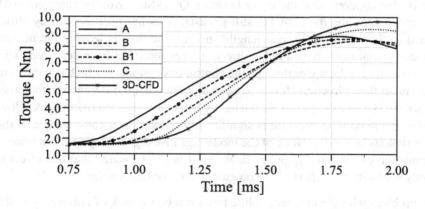

Figure 6.22: Torque trace for versions A - C at 400 Hz pressure pulse

Version D and E differ from C in the effort of reducing the sharp edges at the transition. Figure 6.23 shows that this does not only effect the static pressure loss, but also the transient response. A smoother transition leads to a remarkable increase in maximum pressure.

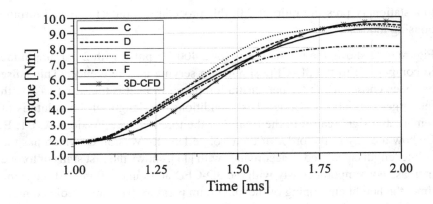

Figure 6.23: Torque trace for Version C - F at 400 Hz pressure pulse

It is also apparent that the delay between QuickSim with the integrated 0D-Turbocharger and the 3D-CFD still persists. This has been previously attributed to the lack of a discrete length; instead, the flow is leaving the turbine housing continuously in a radial direction. Version F has been designed to adopt exactly this behavior through the continuous boundary, which is parallel to the main flow direction. It can be seen that there is no difference to the other versions with a calculated volute length when comparing the rising flank. However, the maximum pressure is significantly reduced as a consequence of the continuous boundary. None of the modelings investigated captures the smooth transition of the arriving pulse; however, it has to be stated that this effect is only seen for the highest of the investigated pulse frequencies.

A problem with the different volute types can be the lack of uniformity in the cells close to boundary. As the 0D-Turbocharger requires a scalar input value, the thermodynamic state variables like pressure, temperature and velocity have to be averaged over the entire boundary face. The deviation between the extreme value and the averaged value are quantified in Table 6.4. For the versions with a pipe extrusion (A, B1, B), this is not a problem, as a fully developed pipe flow is present. Versions C-E show a stronger lack of flow uniformity due to the flow disruption at the transition; however, a deviation below 1.5 % can be acceptable when considering the better transient behavior. For version F, the deviation is so high that the averaging process can be a serious source of error; therefore, this modeling is not recommended.

Table 6.4: Pressure difference in boundary cells of volute outlet

A, B, B1	C	D	E	F
0.2 %	1.4 %	1.2 %	0.8 %	14.1 %

The validation of the 0D-Turbocharger has proven the functionality and robustness of the models implemented for the previously determined operating conditions in depth. Accuracy is significantly improved compared to a standard 1D-Simulation especially for pulses with high pressure gradients. The stability of the modeling is good, and the results relatively independent from the chosen physical time step. Calculation time of the additional 0D-models is negligible compared to the equations solved on the 3D-mesh.

- QuickSim with the 0D-Turbocharger is stable even under the worst case scenarios of a 400 Hz and 2 bar amplitude pressure pulse

- With the 3D-CFD as a reference, the accuracy of QuickSim is significantly better than the industry standard 1D-Simulation, especially for high pressure gradients

- Simulation stability and accuracy do show a dependency on the temporal discretization; however, the typical values from the engine simulation have proven sufficient for the cases investigated.

- Travel length in the turbine housing determines the delay of the pressure pulse as expected, while the shape of the housing influences the pressure gradient and maximum pressure.

7 Application of the 0D-Turbocharger

7.1 Introduction of the Engine and the Operating Point

Excerpts of the application example shown, have also been publish with fewer details in [32]. The modeling, validated on the VHGTB, is now applied to an two-cylinder engine. In order to demonstrate the capabilities of the approach developed, a turbocharged engine has been chosen that is characterized by a strongly pulsating exhaust gas and showing a difficult gas exchange process. The two-cylinder engine has been developed for special applications like jet-skis or snowmobiles and presents a number of characteristics atypical for a series automobile engine. It is usually run under a high load with medium to high rpm, which is reflected in a high bore-to-stroke ratio and a large valve overlap. At the same time, the engine has to be very cost effective, often prohibiting additional features to benefit the low to medium load region.

Table 7.1: Textron MPE 850 four-stroke engine specification

Bore	89 mm
Stroke	68 mm
Displacement	846 ccm
Compression ratio	10.5:1 (modified)
Max speed	7,000 rpm
Valves per cylinder	4
Injection system	DI, lateral
Charging system	Turbocharged, cast exhaust runners including the turbine housing

The two-cylinder engine with a 360° firing interval produces very distinct pulses, especially at low speeds and high load. At lower load, the engine throttles strongly to reduce the mass flow, resulting in interesting scavenging

© Springer Fachmedien Wiesbaden GmbH, part of Springer Nature 2020
A. Kächele, *Turbocharger Integration into Multidimensional Engine Simulations to Enable Transient Load Cases*, Wissenschaftliche Reihe Fahrzeugtechnik Universität Stuttgart, https://doi.org/10.1007/978-3-658-28786-3_7

phenomena due to the large valve overlap. Further information on the engine can be found in Table 7.1 and [32, 60].

As the simulation data will be compared to measurements from the test bench, the simulation domain is extended upon the engine and the turbocharger to include the entire test bench environment. This is comprised of the intake filter, the long pipe for a high-quality mass flow measurement and the charge air intercooler as shown in Figure 7.1. This extension of the simulation domain ensures a good agreement between the simulation and the measurements. The exact replication of the test bench also allows for an accurate positioning of the virtual sensors.

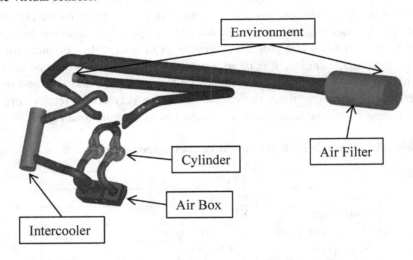

Figure 7.1: MPE-850 simulation domain in test bench setup [32]

In this work, two models of this engine will be compared, differing in the extension of the simulation domain. The smaller model does not have the low pressure part of the intake and exhaust system and consists of 230,000 cells, while the full model shown in Figure 7.1 has 310,000 cells.

As the operating point investigated requires the throttle valve to be partially closed, different implementations of the throttle body are compared. If the butterfly throttle is meshed as a part, the cells needs to be very small to capture the flow phenomena through the narrow openings around the flap. A new mesh has to be generated during the calibration process, either through an adaptive mesh, or, if the changes are significant through an entirely new mesh structure. For transient simulations, this can be a major obstacle as the number of required meshes increases rapidly with a fine temporal resolution or slower load changes.

To minimize this effort, an alternative throttle representation was investigated, in which the throttle is fixed in WOT position and the fluid volume close to the throttle is assigned a porosity causing velocity dependent pressure losses in the respective cells. Both versions are shown in Figure 7.2.

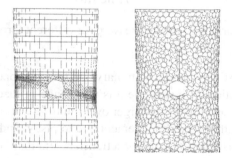

Figure 7.2: Mesh representation (left) and porous representation (right)

For the operating point used later in this chapter, both implementations have been realized to investigate the differences in the wave reflection between the two representations. In the case of a strongly closed throttle flap, the waves are reflected, typical for a 'closed end'. Due to the definition of the porous volume, pressure loss is distributed isotropically over all cells, unable to represent a distinct 'end' for the wave to be reflected at.

Figure 7.3 shows, that the frequency of the pressure oscillations is not affected and that the amplitude differences are very small. It can be concluded, that for the case investigated a throttle representation via a porous volume is valid and has very little impact on the results. Additionally, the size of the porous volume does not have a significant impact on the pressure propagation.

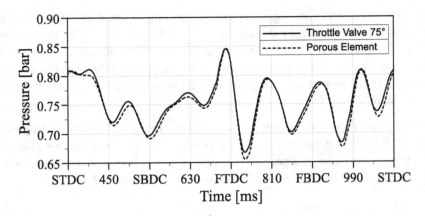

Figure 7.3: Comparison of intake pressure for different throttle models

The engine is investigated under stationary and transient operating conditions. In this context, stationary means there is no change in load or speed for the combustion engine and the turbocharger over multiple working cycles. It must not be confused with the mesh movement due to piston and valve motion. In contrast to that, it is characteristic for a transient simulation to include a high number of consecutive working cycles and load or speed changes. Usually transient simulations are performed to investigate the impact of a certain event, e.g an increased load requirement by the driver.

The chosen engine operating point will be used during a stationary analysis and is the starting point for the following transient simulation of a load change.

Table 7.2: Operating point of the two-cylinder engine

Speed	3,000 rpm
Throttle position	15° open
p22	0.77 bar
p3	1.14 bar
Lambda	1.0
Fuel	RON 95

At 3,000 rpm and a low load (15° opening angle of throttle valve), the turbine operates in an acceptable area of the performance map, while the compressor is close to or even beyond the surge line. The high pressure ratio of the compressor is a consequence of the almost closed throttle valve limiting the mass flow of compressed air.

7.2 Calibration of the Engine Models

In the following investigations, measurements from the test bench will be compared to a 1D-Simulation used and calibrated in [26] and to the 3D-CFD Quick-Sim with the 0D-Turbocharger. In a first step, both models are calibrated to the air mass flow and power output from the test bench, the results of which are displayed in Table 7.3. The calibration is necessary for full engine models as the simulation can only be a simplified reproduction of a complex physical process. The models used have to be adapted with the case-specific parameters.The calibration has to also compensate for inaccuracies in the input data like valve lash, injector targeting, liquid and vaporous fuel properties, etc.

Table 7.3: Two-cylinder engine stationary results after the calibration process

	Air kg/h	Fuel kg/h	TC speed rpm	IMEP bar	Airbox pressure bar
Test bench	53.0	4.8	66,940	6.68	0.77
1D-Simulation	52.5	3.64	66,100	6.90	0.79
QuickSim	52.6	4.95	66,250	6.90	0.76

It can be observed that the largest deviation from the test bench data is the fuel flow of the 1D-Simulation. The required mass flow is substantially underestimated because the fuel-air mixture is actually pushed back into the intake runners shortly before the intake valves close. In the next cycle, this mixture is directly scavenged through the cylinder and into the exhaust system due to the large valve overlap. The 1D-Simulation does not predict this behavior, because it does not have spatial information about the the fuel in the cylinder.

The calibration procedure for QuickSim is intentionally very similar to setting an operating point at the test bench and started at a very high turbocharger inertia, fixing the shaft speed and temporarily decoupling the exhaust from the intake side. With an acceptable combustion ensured, the intake side, especially air and fuel mass flow, are calibrated in addition to the air box and low pressure indication. With the correct mass flow and cylinder charge, the combustion can be calibrated, strongly influencing the turbine enthalpy. In the next step, the turbine performance is calibrated in order to comply with the pressure measured on the test bench and balancing the shaft torque equilibrium. Finally the shaft inertia is reduced down to its physical value enabling the shaft to change speed. In addition to the values applicable on the test bench, additional model constants like the diameter of the spark kernel or the flame wrinkling due to turbulence have to be calibrated. A detailed methodology can be found in Appendix A1.1. The high number of calibration steps emphasizes the value of a good initial guess and the requirement of a fast calculation; otherwise, this task can not be performed in an acceptable time.

7.3 Stationary Analysis: Part Load Operating Point

The chosen operating conditions are in a low part load and the cylinder scavenging process is heavily influenced by the valve timing shown in Figure 7.4 namely by the large valve overlap and the late IVC.

Figure 7.5 shows the pressure measured at the intake runners close to the cylinder head. As already shown in Table 7.3 average pressure of the simulations differs slightly from the measurement on the test bench. However, the differences are quite small, especially for QuickSim. The differences in oscillation frequency are of higher significance. While QuickSim predicts the results of the test bench very well, the 1D-Simulation calculates a significantly shorter period. This is likely a result of the very complex runner and air box geometry which is very difficult to model in a 1D environment.

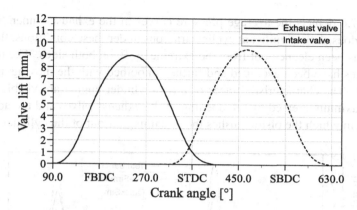

Figure 7.4: Valve lift of the two-cylinder engine

Both simulations share a reduced amplitude compared to the test bench. The root cause for this behavior can not be determined, however multiple reasons can contribute to it. One plausible explanation, besides measurement inaccuracies, can be found in the physically oscillating air box walls. Due to the low air box pressure, the plastic walls are deflected, possibly amplifying the oscillations.

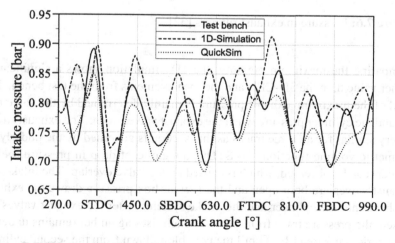

Figure 7.5: Pressure in the intake runner [32]

Figure 7.6 depicts the average pressure curves in the exhaust runner. As the engine does show significant cyclic variations under these conditions, the data basis has been cleaned out from the extreme 20 % on both sides. The remaining traces have been averaged and further smoothed with the moving average method. The characteristic traces of a two-cylinder engine are visible in the high maximum created by the opening of the exhaust valves and the adjacent plateau in which the piston pushes the remaining air out of the cylinder.

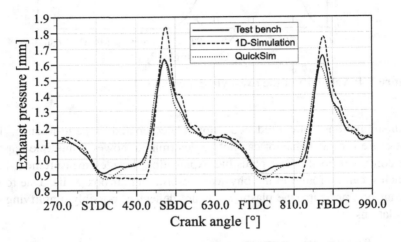

Figure 7.6: Pressure in exhaust runners [32]

Comparing the maximum pressure, the 1D-Simulation shows a significantly higher value than the 3D-CFD and the measurements from the test bench. The level of the maximum exhaust pressure is mainly determined by the valve timing and the cylinder pressure when the exhaust valves open. The exhaust stroke is very similar for all three traces as this section is governed by the underlying geometric volume function. At STDC, a significant drop in pressure (below 0.9 bar) can be observed, which is caused by the valve overlap. The intake and exhaust valves are both open and the low air box pressure draws the exhaust gas back into the cylinder and the intake system. After the exhaust valves are closed, the pressure trace from the test bench rises again but remains under atmospheric pressure (1 bar) until the next blow-down from the second cylinder is apparent. This pressure increase can only be explained by a reversed mass flow through the turbine.

In QuickSim, the drop is a little stronger than on the test bench, but the turbine model allows for a reversed mass flow through the turbine enabling the pressure to catch up to the measurement. For demonstration purposes, reversed mass flow has not been allowed in the 1D-Simulation, forcing the pressure to remain at the lowest level of the drop. Information on reversed mass flow is rarely available from the manufacturer. To counteract this problem, the model introduced in Chapter 5.2.2 was developed.

Figure 7.7 has the measured turbine performance map drawn in continuous lines and the operating points of the turbine during one engine cycle as markers. The algorithm for generating the performance maps extrapolates the individual speed lines, to meet at zero mass flow and a pressure ratio of one. It has already been demonstrated in Chapter 5.2.2 that this behavior is not realistic and that the intersection should depend on the shaft speed and be between a pressure ratio of 1.0 and 1.03. The model implemented ensures this and controls the mass flow through the turbine for pressure ratios below 1.05. To encounter the unknown behavior of the turbocharger, the model can be calibrated by two parameters.

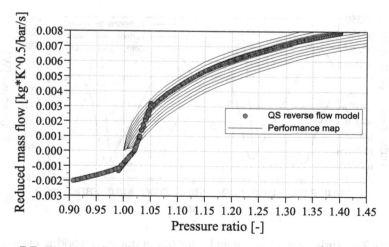

Figure 7.7: Reversed mass flow in turbine during one working cycle [32]

Due to the changing conditions generated by the combustion engine, the turbocharger operates in an area (often refereed to as 'hysteresis loop') of the performance map, rather than an operating point. This is also reflected by the turbocharger shaft speed in Figure 7.8, which varies by approximately 400 rpm within a working cycle with an inertia of 1.7E-5 $kg * m^2$ for the rotor assembly. The shaft speed is driven by the power difference of the turbine and compressor wheel and the mechanical losses in the bearings. In the case presented here the thermo-mechanical map includes the losses into the isentropic turbine efficiency. The compressor power is almost constant, as the variance in mass flow and pressure ratio is small. Most of the time, pulsations in the intake systems are weaker than in the exhaust system due to a smaller pressure difference when the valve opens and the large volume of charge air cooler, air box and runners.

For better readability, the turbine power is displayed as absolute value, as it would otherwise have an inverted sign. It is close to zero for one third of the working cycle, because the mass flow that delivers the enthalpy to the turbine is only available when the exhaust valves are opened. In this analysis the power during the reversed flow phase is set to zero but can of course be adapted to a negative value to account for friction, etc.

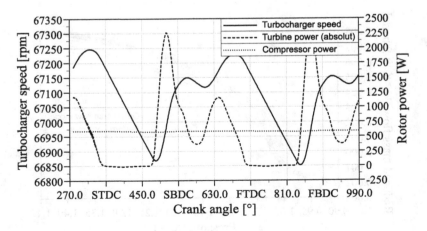

Figure 7.8: Turbocharger speed and rotor power during a working cycle [32]

A comparison of the exhaust pressure traces between a simulation with and without the 0D-Turbocharger is shown in Figure 7.9. Without using the 0D-Turbocharger, the simulation domain ends at the turbine housing inlet, where time dependent pressure and temperature boundary conditions from an external 1D-Simulation have to be defined. These boundary conditions are the target value for the simulation controls and shown as a continuous line, while the actual pressure is shown as dashed line. If the simulation is stable, the two lines should lie on top of each other. A difference between them is an indicator of instability. After STDC, the 0D-Turbocharger is able to refill the exhaust runners by reversing the mass flow. The boundary conditions, however, have been derived from a 1D-Simulation that does not allow the reversed mass flow and drives QuickSim (a presumably more accurate 3D-CFD simulation) to deliver bad results.

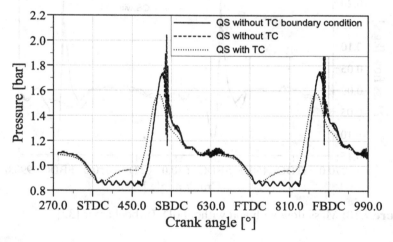

Figure 7.9: Pressure traces with and without 0D-Turbocharger [32]

This is an extreme example of how the boundary conditions' quality can influence the simulation quality; however, also smaller errors deteriorate the latter. The next difference is the maximum pressure peak shortly before SBDC. Obviously, the simulation with boundary conditions is again influenced by the wrong input. Shortly after the peak, the target and the real pressure deviate dramatically leading to heavy oscillations.

The explanation can be found in the mass flow plot in Figure 7.10, where the flow entering the system is displayed with a positive sign and the flow leaving the system with a negative sign. Without the 0D-Turbocharger, the simulation is forced to push a strong mass flow into the simulation domain before SBDC, in order to follow the target pressure. As the pressure peak is reached, the mass flow has to be reduced drastically, resulting in heavy oscillations. This causes the stability issues described here, requiring a high number of inner iterations per time step or causing the simulation to diverge. For the two-cylinder engine chosen here, this behavior is extremely critical as the large valve overlap connects the intake system with the exhaust system allowing the errors introduced on the exhaust side to affect the intake side.

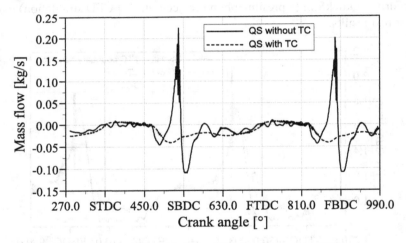

Figure 7.10: Mass flow with and without 0D-Turbocharger [32]

In the application example of the 0D-Turbocharger modeling approach, four main benefits have been presented so far:

- The turbocharger enables the simulation domain to be extended (exhaust gas treatment, post-oxidation ...)

- The system has the flexibility to adapt to changing parameters like valve timing, etc., without the need for new boundary conditions

- Higher stability of the included turbocharger enables the code to run with larger physical time steps

- With an integrated turbocharger, the influence of the boundary conditions on the combustion and gas exchange process is minimized

7.4 Transient Analysis: Load Change via Ignition Retarding

Before the transient analysis, a detailed optimization of the simulation parameters was conducted to improve the simulation speed without affecting result quality. A significant improvement in calculation time is achieved by the increased stability of the simulation, allowing an increased physical time step. As shown in Table 7.4 a doubling of the time step does not lead to an inversely proportional decrease in time as the number of required inner iterations increases. Other fields of improvement are the liquid fuel representation via droplets and a modified valve opening strategy. As the simulation produces a remarkable amount of data for the large number of working cycles, what is required for future post-processing or for a restart has to be carefully selected.

The values found are certainly not the optimum for every engine, but might have to be readjusted individually.

Table 7.4: Optimization of calculation time for one working cycle in full load

Before optimization	8:23 h
Increase physical time step from 0.5° CA to 1° CA	5:48 h
Increase droplet per parcel ratio by 100 %	5:42 h
Improved valve opening strategy	5:40 h
Improved data storage	5:30 h

With the increasing relevance of RDE, the development focus has shifted towards transient engine behaviour, especially critical for turbocharged engines. The behavior of the engine and the turbocharger during a transient maneuver cannot be captured by stationary simulations only. In fact, the imbalances (for example, turbocharger shaft speed does not match the throttle position) are the driver of the transient maneuver. It is also very difficult to start a simulation containing only the later part of a transient maneuver as the starting conditions are a result of a transient history. This has led other authors to simulate the entire run-up of the turbocharger shaft from a standstill [7].

In this chapter, a change in load which is caused by an abrupt retarding of the ignition is simulated and analyzed. Before the transient simulation, two stationary simulations have to be carried out; one for the starting point, and one for the equilibrium after the maneuver. The latter is only required, if the change is so drastic, that the model calibration of the starting point cannot be maintained. Examples for this can be wall temperatures (if they are fixed and not calculated), or combustion settings, if the combustion is heavily altered.

If the calibration values need to be changed during the simulation, a law has to be defined according to which the modification has to be performed. This is also the case if the throttle or waste gate position is altered, or other input values like the fuel mass flow or the injection timing are changed. If the data are to be compared to the test bench, it is critical to obtain accurate input data, because the employed controls can overshoot and/or undershoot the target value remarkably. To accurately simulate the injected fuel mass and, consequently, the combustion lambda, it is beneficial to record the injection timing for each working cycle.

The starting point of the transient maneuver is the operating point closely examined in Chapter 7.3. Ignition timing is retarded by 20°, from 696° CA to 716° CA, leading to a shift in center of combustion (50 % fuel burned) from 10° to 32° after FTDC. The retarded combustion is less efficient, reducing the work transferred to the piston and hence increasing the exhaust gas enthalpy. The higher enthalpy available to the turbine changes the operating point and increases the shaft speed, leading to a higher mass flow and pressure ratio of the compressor. To maintain a stoichiometric air/fuel ratio, fuel mass flow needs to be increased, again changing the enthalpy available at the turbine. The number of working cycles required to reach an equilibrium depends strongly on the engine configuration ranging up to one hundred or more. In this example, the retarded ignition reduces the engine brake power by 25 %, increasing the average cylinder temperature when the exhaust valve opens from 1,500 K to 1,900 K.

Figure 7.11 shows the turbocharger speed during the transient maneuver. The ignition is retarded in cycle zero, leading to a steady increase in turbocharger speed until a new equilibrium is found in approximately cycle 80. While the overall agreement between QuickSim with the 0D-Turbocharger and the measurement on the test bench is good, there are some areas of interest which require a closer look.

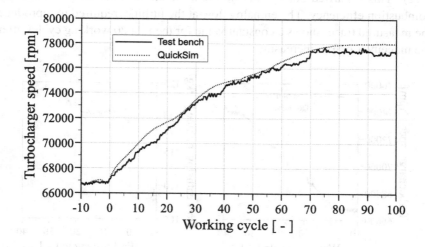

Figure 7.11: Turbocharger speed after load change [32]

The speed trace from the simulation does show a slight oscillation, which is assumed to be caused by the deviation from the target cylinder lambda, replicated from the test bench measurement. Due to the higher cyclic variations and the therefore required smoothing of the measurement trace, it is unclear whether this behavior is also detected on the test bench.

The speed traces are almost identical for the first two working cycles, but after those, the simulation predicts a significantly faster increase until cycle 25. The reason for this can be found in the fixed value of the cylinder and exhaust runner wall temperature in the simulation. They have been calibrated to be in the middle of the expected range between the starting and equilibrium points. Shortly after the transient event, the rough inner surface and the high thermal inertia of the cast exhaust runners extract a lot of enthalpy from the exhaust gas before it reaches the turbine, thus delaying the turbocharger spin-up. After the surface temperature is increased, the difference between the gas and the wall and, consequently, the heat flux,is reduced.

To substantiate this hypothesis, the turbocharger shaft speed is controlled to follow the measured trace as closely as possible in a new simulation (see Figure 7.12). This is carried out by altering the available flow enthalpy through the combustion efficiency. The enthalpy flow at the turbine, required to reproduce the measured trace, shows a constant value for the first 20 working cycles after the maneuver before increasing.

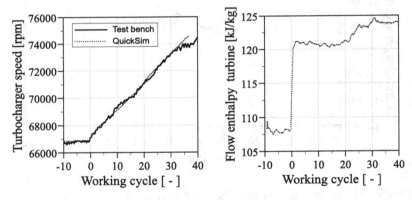

Figure 7.12: Turbine flow enthalpy for test bench shaft speed [32]

One possible explanation is that the combustion in the original simulation (see Figure 7.11) is different than at the test bench, which can be ruled out through the cylinder pressure sensor. Consequently, the wall temperature and housing heat capacity cause this phenomena.

The pressure sensor in the intake system (close to the intake channels) detects a constant pressure increase on the test bench and in the simulation until the new equilibrium is reached as depicted in Figure 7.13. The simulation shows a lower value at the beginning and the end of the maneuver. Shortly after cycle zero, the pressure increases faster for the simulation, which is in accordance with the turbocharger speed.

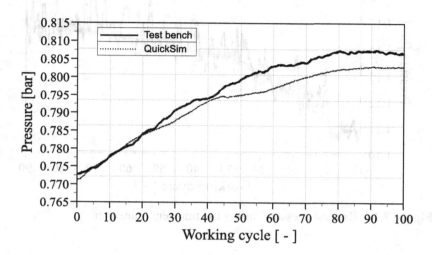

Figure 7.13: Intake pressure during the transient maneuver

The smoothed exhaust pressure traces of the transient maneuver can be seen in Figure 7.14. In contrast to the intake side, a sharp increase after retarding the ignition is detected on the test-bench and in the simulation. After this initial jump, the pressure rises slowly as does the boost pressure and mass flow through the engine.

The mass flow through the engine is presented in Figure 7.15. Due to the inertia of the air mass and the sensor, a sharp peak after the triggering event is not evident in the trace of the test bench, however the simulation clearly shows this behavior. Interestingly, the simulation and test bench are in very good agreement when equilibrium is reached, despite the difference in intake pressure.

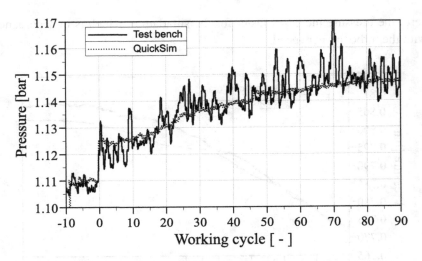

Figure 7.14: Exhaust pressure during the transient maneuver

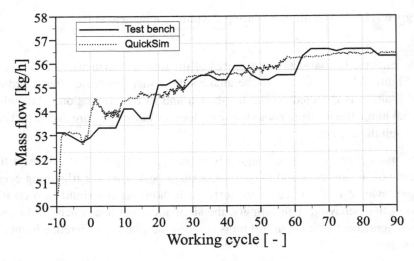

Figure 7.15: Mass flow during the transient maneuver

The integration of a 0D-Turbocharger into the engine environment of Quick-Sim has proven to ensure a very stable and fast simulation. For highly accurate results, reliable test bench data and performance maps of high quality are required in order to enable a good calibration.

The average calculation time for the maneuver demonstrated here is 6:50 h per cycle and higher than Table 7.4 indicates. For 100 consecutive working cycles, this adds up to 28 days on a single core (Intel Xenon E5-2643 v4). The part load operating point, which is less numerically stable and requires a higher number of inner iterations per time step, contributed to this. During the transient run, only carefully selected data is stored for every iteration; however, over the course of the simulation, these data files grow tremendously in size. This is especially apparent for the droplet number, as each fuel droplet is assigned a unique number that will not be used twice during this simulation. These numbers have to be stored and made available for post processing.

The following conclusions can be drawn from the transient analysis of the two-cylinder engine:

- With the 0D-Turbocharger, the simulation of a transient load change over 120 consecutive working cycles is possible with good agreement to the test bench

- Due to the long simulation duration, it is beneficial to invest into an optimization and acceleration of the simulated model

- Attention has to be paid to the controls adjusting injection timing, fuel mass flow, wall temperature, etc., during the simulation

8 Conclusion and Outlook

The aim of the present work is to enable a transient 3D-CFD simulation over multiple working cycles for turbocharged internal combustion engines. In this context, a transient simulation is defined as a change in the engine operating point, e.g. speed or load. Due to the strong interaction between a turbocharger and a combustion engine, an integrated approach through a coupled engine-turbocharger simulation is required to ensure a global system response. The turbocharger has to viewed as a second link, connecting the intake and the exhaust manifolds, with the valve overlap being the first one.

Different modeling approaches have been evaluated under the premise of being used in virtual engine development. The integrated 0D-Turbocharger has been found to be the one best suited, due to the fast calculation time, the acceptable level of spatial resolution and the high availability of the performance maps as input data. To ensure a broad application, the modeling has been extended to work with both, aerodynamic and thermo-mechanical performance maps. Further extension of the operating range has been achieved by additional models ensuring the required stability during surge or very low and reversed mass flow. The surge modeling of the pressure ratio has been proven to work best with an exponential modeling, while the efficiency can be modeled linearly. Reversed mass flow can be handled either via an extended performance map, generated during pre-processing, or through a newly developed model for fast calibration during the simulation.

The validation of the modeling is carried out on a virtual hot gas test bench, designed to operate under a stationary and pulsating flow. Recreating the maximum pressure gradient from multiple examples, an artificial pulse is found to replicate the worst case in the shape of a sinusoidal function with 400 Hz and two bar amplitude. The comparison of the system reaction between three simulation environments shows that the chosen modeling is a significant improvement over a conventional 1D-Simulation. However, especially for the pressure gradients in the worst case scenario, the difference between the full 3D-CFD, serving as a reference, and the 0D-Turbocharger in QuickSim can

© Springer Fachmedien Wiesbaden GmbH, part of Springer Nature 2020
A. Kächele, *Turbocharger Integration into Multidimensional Engine Simulations to Enable Transient Load Cases*, Wissenschaftliche Reihe Fahrzeugtechnik Universität Stuttgart, https://doi.org/10.1007/978-3-658-28786-3_8

be observed. Simulation stability is ensured for the range of pressure pulse gradients demanded and additional investigations show the differences for alternative turbine housing representations.

No optimal physical time step size could be determined; however, a clear tendency is observed, showing that smaller time steps slightly delay the pressure increase at the turbine rotor in conjunction with a minor reduction in peak pressure. As the differences are relatively small compared to other influences, it is recommended to chose the time step in a way that is suited to save computational effort for the internal combustion engine. Increasing engine speed, often coinciding with a higher pressure gradient, will therefore automatically lead to a shorter time step.

The validated integrated 0D-Turbocharger is applied to a two-cylinder engine, demonstrating the supremacy in boundary condition quality compared to a noncoupled simulation. The higher quality allows for short distances between the exhaust valves and the turbine without affecting the simulation stability. The 0D-Turbocharger also removes constraints to the gas exchange imposed by fixed boundary conditions, leading to a higher accuracy, for example, in the case of reverse mass flow or surge.

The same turbocharged engine is used to simulate the transient behavior during a change in engine load through ignition retarding. The system reaction of the engine and turbocharger is simulated over more than 100 consecutive working cycles which correspond to four seconds in physical time. The overall agreement with the measurements from the test bench is very good.

In both the stationary and transient simulation, the importance of calibration has to be emphasized. The 0D-Turbocharger introduces a large number of new parameters to the simulation and the link of intake and exhaust manifold via the shaft speed increase the complexity remarkably. A procedure is defined to standardize and accelerate the calibration process; however, the importance of a good initial guess derived, for example, from test bench data, cannot be stressed enough.

A possibility for future development is the implementation of heat transfer into the 0D-Turbocharger, which is especially important at low speeds and loads.

This can be done in pre-processing when generating the performance map or during simulation via additional modeling.

As shown for the two-cylinder engine, it is important for the transient analysis to use self adjusting wall temperatures, instead of a fixed ones, as it does affect the spin-up of the shaft considerably.

Another possible improvement concerns the modeling of the flow structure after the turbine. For the results presented here a flow without swirl has been assumed as it does not have a large impact on the performed analysis. If the exhaust gas after-treatment is of interest, this swirl has to be considered, as it determines the incident flow of the catalyst or the urea injector. The swirl can be calculated with sufficient accuracy from the mass flow, the shaft speed and the geometric turbine blade angle.

Bibliography

[1] J. Anderson. *Computational Fluid Dynamics.* McGraw-Hill Education - Europe, 1995.

[2] R. Aymanns, J. Scharf, T. Uhlmann, and D. Lückmann. A Revision of Quasi Steady Modelling of Turbocharger Turbines in the Simulation of Pulse Charged Engines. *16. Aufladetechnische Konferenz Tagungsband,* 2011.

[3] N. C. Baines, A. Hajilouy-Benisi, and J. H. Yeo. The Pulse Flow Performance and Modelling of Radial Inflow Turbines. *Proceedings 5th International Conference on Turbochargers and Turbocharging of the IMechE, Paper C484/006/94,* 1994.

[4] M. Bargende. Berechnung und Analyse innermotorischer Vorgänge, lecture notes, 2013.

[5] M. Bargende. *Ein Gleichungsansatz zur Berechnung der instationären Wandwärmeverluste im Hochdruckteil von Ottomotoren.* PhD thesis, Techn. Hochschule Darmstadt, 1991.

[6] R. Benson. Nonsteady Flow in a Turbocharger Nozzleless Radial Gas Turbine. *SAE Technical Paper 740739,* 1974.

[7] B. Boose. *3D CFD Simulation von Turboladern innerhalb einer Motorumgebung.* PhD thesis, University of Stuttgart, 2014.

[8] T. Cao, L. Xu, M. Yang, and R. F. Martinez-Botas. Radial Turbine Rotor Response To Pulsating Inlet Flow. *Proc. of ASME Turbo Expo 2013: Power for Land, Sea and Air,* GT2013-95182, 2013.

[9] M. Capobianco and S. Marelli. Transient Performance of Automotive Turbochargers: Test Facility and Preliminary Experimental Analysis. *SAE Technical Paper Series, 2005-24-066,* 2005.

© Springer Fachmedien Wiesbaden GmbH, part of Springer Nature 2020
A. Kächele, *Turbocharger Integration into Multidimensional Engine Simulations to Enable Transient Load Cases,* Wissenschaftliche Reihe Fahrzeugtechnik Universität Stuttgart, https://doi.org/10.1007/978-3-658-28786-3

[10] M. V. Casey and T. M. Fesich. The Efficiency of Turbocharger Compressors With Diabatic Flows. *Journal of Engineering for Gas Turbines and Power*, 2010.

[11] H. Chen. Steady and Unsteady Performance Of Vaneless Casing Radial-Inflow Turbines. *Dissertation, University of Manchester*, 1990.

[12] H. Chen, I. Hakeem, and R. F. Martinez-Botas. Modelling of a Turbocharger Turbine under Pulsating Inlet Conditions. *Proceedings of the IMechE, Part A: Journal of Power and Energy*, vol. 210(no. 5):pp. 397–408, 1996.

[13] H. Chen and D. Winterbone. A Method to Predict Performance of Vaneless Radial Turbines Under Steady and Unsteady Flow Conditions. *IMechE Turbocharging and Turbochargers,*, (C405/008):13–22, 1990.

[14] H. Chen and D. E. Winterbone. A one-dimensional performance model for turbocharger turbine under pulsating inlet condition. *Proceedings 11th International Conference on Turbochargers and Turbocharging of the IMechE*, 2014.

[15] M. Chiodi. *An Innovative 3D-CFD-Approach towards Virtual Development of Internal Combustion Engines*. PhD thesis, University of Stuttgart, 2010.

[16] M. Chiodi, A. Kaechele, M. Bargende, D. Wichelhaus, and C. Poetsch. Development of an Innovative Combustion Process: Spark-Assisted Compression Ignition. *SAE International Journal of Engines*, 10(5), 2017.

[17] C. Copeland, P. Newton, R. F. Martinez-Botas, and M. Seiler. A Comparison of Timescales Within a Pulsed Flow Turbocharger Turbine. *IMechE 10th International Turbochargers and Turbocharging*, pages 389–404, 2012.

[18] A. Costall, S. Szymko, R. F. Martinez-Botas, D. Filsinger, and D. Ninkovic. Assesment of Unsteady Behaviour in Turbocharger Turbines. *Turbo Expo: Power for Land, Sea, and Air*, Volume 6: Turbomachinery, 2006.

[19] S. L. Dixon and C. A. Hall. *Fluid Mechanics and Thermodynamics of Turbomachinery.* Butterworth-Heinemann/Elsevier, 2010.

[20] J. El Hadef, G. Colin, V. Talon, and Y. Chamaillard. New Physics Based Turbocharger Data Maps Extrapolation Algorithms: Validation on a Spark-Ignited Engine. In *2012 IFAC Workshop on Engine and Power-train Control, Simulation and Modeling (ECOSM)*, 2012.

[21] C. A. J. Fletcher. *Computational Techniques for Fluid Dynamics 1.* Springer Berlin Heidelberg, 1998.

[22] C. Fredriksson and N. Bains. The mixed flow forward swept turbine for next generation turbocharged downsized automotive engines. *ASME GT2010-23366*, 2010.

[23] Gamma Technologies. GTISE Help. *(English)*, 2017.

[24] R. Golloch. *Downsizing bei Verbrennungsmotoren: Ein wirkungsvolles Konzept zur Kraftstoffverbrauchssenkung (VDI-Buch) (German Edition).* Springer, 2005.

[25] M. Grill, M. Bargende, D. Rether, and A. Schmid. Quasi-dimensional and Empirical Modeling of Compression-Ignition Engine Combustion and Emissions. In *SAE Technical Paper Series.* SAE International, apr 2010.

[26] T. Günther, M. Grill, A. Kächele, M. Chiodi, J. Eder, H. Berner, and M. Bargende. Nachoxidation zur Drehmomentsteigerung und Erfüllung der Emissionen während des spülenden Ladungswechsels am turboaufgeladenen Ottomotor. *Aufladetechnische Konferenz, Dresden*, 2017.

[27] J. Heywood. *Internal Combustion Engine Fundmentals.* McGraw-Hill Education, 1988.

[28] H. Hireth and P. Prenninger. *Aufladung der Verbrennungskraftmaschine - Der Fahrzeugantrieb.* Springer Verlag Wien, 2003.

[29] G. Hohenberg. Advanced Approaches for Heat Transfer Calculations. *SAE Tech Series*, (Paper 790825), 1979.

[30] International Organization for Standardization. ISO 5389:2005 Turbocompressors Performance test code. Technical report, ISO, 2005.

[31] J.-P. Jensen, A. Kristensen, S. Sorenson, N. Houbak, and E. Hendricks. Mean Value Modeling of a Small Turbocharged Diesel Engine. In *SAE Technical Paper Series*. SAE International, feb 1991.

[32] A. Kaechele, M. Chiodi, and M. Bargende. Virtual Full Engine Development: 3D–CFD Simulations of Turbocharged Engines under Transient Load Conditions. *SAE Technical Paper 2018-01-0170*, 2018.

[33] E. Laurien and H. Oertel. *Numerische Strömungsmechanik*. Springer Fachmedien Wiesbaden, 2013.

[34] J. Lotterman, N. Schorn, D. Jeckel, F. Brinkmann, and K. Bauer. New Turbocharger Concept for Boosted Gasoline Engines. *16. Aufladetechnische Konferenz*, 2011.

[35] B. Lüddecke. Comprehensive Study on Insulation Influences. *Internal Report, IHI Charging Systems International GmbH*, May 2012.

[36] B. Lüddecke. *Stationäres und instationäres Betriebsverhalten von Abgasturboladern*. Springer Fachmedien Wiesbaden, 2016.

[37] G. P. Merker and R. Teichmann. *Grundlagen Verbrennungsmotoren*. Springer Fachmedien Wiesbaden, 2014.

[38] N. Mizumachi, H. Yoshiki, and T. Endoh. *A Study on Performance of Radial Turbine Under Unsteady Flow Conditions*. Institute of Industrial Science, University of Tokyo, 1979.

[39] B. Noll. *Numerische Strömungsmechanik*. Springer Berlin Heidelberg, 1993.

[40] D. Palfreyman and R. Martinez-Botas. The Pulsating Flow Field in a Mixed Flow Turbocharger Turbine: An Experimental and Computational Study. *ASME, Journal of Turbomachinery*, Vol 127((1)):144– 155, 2005.

[41] F. Pischinger and A. Wünsche. The Characteristic Behavior of Radial Turbines and its Influence on the Turbocharging Process. In *Proceedings CIMACCongress*, Tokyo, 1977.

[42] S. Rajoo and R. F. Martinez-Botas. Unsteady Effects in a Nozzled Turbocharger Turbine. *Journal of Turbomachinery*, 132.3(p 031001):26,29,31, 2010.

[43] D. Rether, M. Grill, A. Schmid, and M. Bargende. Quasi-Dimensional Modeling of CI-Combustion with Multiple Pilot- and Post Injections. *SAE International Journal of Engines*, 3(1):12–27, apr 2010.

[44] A. Romagnoli, R. F. Martinez-Botas, and S. Rajoo. Turbine Performance Studies for Automotive Turbochargers. *Proceedings 9th International Conference on Turbochargers and Turbocharging of the IMechE*, 2010.

[45] J. S. Scharf. *Extended Turbocharger Mapping and Engine Simulation*. PhD thesis, Techn. Hochschule Aachen, 2010.

[46] A. Schmid, M. Grill, H. J. Berner, M. Bargende, S. Rossa, and M. Böttcher. Development of a Quasi-Dimensional Combustion Model for Stratified SI Engines. *SAE Paper 2009-01-2659*, 2009.

[47] S. Schneider, M. Chiodi, H. Friedrich, and M. Bargende. Development and Experimental Investigation of a Two-Stroke Opposed-Piston Free-Piston Engine. In *SAE Technical Paper Series*. SAE International, nov 2016.

[48] N. Schorn, V. Smiljanovski, U. Späder, R. Stalman, and K. H. Turbocharger Turbines in Engine Cycle Simulation. *13. Aufladetechnische Konferenz*, 2008.

[49] Society of Automotive Engineers. Turbocharger Gas Stand Test Code, SAE J1826 APR89. *SAE Recommended Practice*, 1989.

[50] S. Szymko, R. F. Martinez-Botas, and K. R. Pullen. Experimantal Evaluation of Turbocharger Turbine Performance Under Pulsating Folw Conditions. *ASME Paper No GT2005-68878*, 2005.

[51] The American Society of Mechanical Engineers (ASME). PTC 10-1997 Performance Test Code on Compressors and Exhausters. Technical report, ASME, 1977.

[52] The Association of German Engineers (VDI). VDI 2045-1, Acceptance and Performance Tests on Turbo Compressors and Displacement Compressors (1993) — Part 1: Test Procedure and Comparison with Guaranteed Values. Technical report, VDI, 1993.

[53] Transparency Market Research. Automotive Turbochargers Market (Vehicles - HCV, LCV, Passenger Vehicles, Sports Cars, and Off-highway Vehicles; Technologies - Twin Turbo, Variable Geometry Technology, and Wastegate; Fuel - Gasoline and Diesel; End Users - OEMs and Aftermarket) - Global Industry Analysis, Size, Share, Growth, Trends, and Forecast 2016 - 2024. *Report TMC*, 2017.

[54] T. Uhlmann, B. Höpke, J. Scharf, D. Lückmann, R. Aymanns, K. Deppenkemper, and H. Rohs. Best-In-class Turbochargers for Best-In-Class Engines? A Quantification of Component Design Parameter Impact on Engine Performance. *Aufladetechnische Konferenz Dresden*, 2012.

[55] H. Versteeg and W. Malalasekra. *An Introduction to Computational Fluid Dynamics: The Finite Volumen Method*. Prentice Hall, 2007.

[56] F. J. Wallace and J. Adgey. Paper 1: Theoretical Assessment of the Non-Steady Flow Performance of Inward Radial Flow Turbines. *Proceedings of the Institution of Mechanical Engineers, Conference Proceedings, September 1967*, Vol. 182(No. 8 22-36), 1967.

[57] F. J. Wallace, J. M. Adgey, and G. P. Blair. Performance of Inward Radial Flow Turbines under Non-Steady Flow Conditions. *Proceedings of the Institution of Mechanical Engineers*, Vol 184(No. 1):p. 183–196, 1969.

[58] J. Warnatz, U. Mass, and R. Dibble. *Combustion*. Springer Berlin Heidelberg, 2006.

[59] J. F. Wendt. *Computational Fluid Dynamics*. Springer Berlin Heidelberg, 2009.

[60] M. Wentsch. *Analysis of Injection Processes in an Innovative 3D-CFD Tool for the Simulation of Internal Combustion Engines*. PhD thesis, University of Stuttgart, 2018.

[61] M. Wentsch, A. Perrone, M. Chiodi, M. Bargende, and D. Wichelhaus. Enhanced Investigations of High-Performance SI-Engines by Means of 3D-CFD Simulations. *SAE Tech Paper 2015-24-2469*, 2015.

[62] D. Winterbone, B. Nikpour, and G. Alexander. Measurement of the Performance of a Radial Inflow Turbine in Conditional Steady and Unsteady Flow. In *4th Proceedings Institution of Mechanical Engineering Conference on Turbocharging and Turbochargers, IMeshE-Paper C405/05*, pages 153–162, London, 1990.

[63] D. E. Winterbone and R. J. . Pearson. *Design Techniques for Engine Manifolds: Wave Action Methods for IC Engines*. Professional Engineering Publishing, 1999.

[64] G. Woschni. A Universally Applicable Equation for the Instantaneous Heat Transfer Coefficient in the Internal Combustion Engine. *SAE Tech Paper Series*, (Paper 670931), 1967.

[65] K. Zinner. *Aufladung von Verbrennungsmotoren: Grundlage - Berechnung - Ausführungen*. Osten, W., 2013.

[66] A. Zuarini. Numerical Analysis of Combustion Process in Four-Strokes Spark Ignition Engines. Master's thesis, Diplomarbeit IVK Universität Stuttgart and Politecnico di Torino, 2002.

Appendix

A.1 Appendix 1

Table A1.1: Models chosen in STAR-CCM+ for the calculation of the aerodynamic turbine performance map

Working fluid	Simplified, ideal gas with the properties of nitrogen specific heat according to exhaust gas caloric table of QuickSim
Turbulence modeling	Two-Layer k-ε model with 'All y+ Wall Treatment'
Solver	Coupled Flow and Energy with 2nd order implicit method
Rotation	Relative Reference Frame
Time modeling	Stationary

Table A1.2: Models chosen in STAR-CCM+ for the calculation of pressure pulses on the VHGTB

Working fluid	Simplified, ideal gas with the properties of nitrogen specific heat according to exhaust gas caloric table of QuickSim
Turbulence modeling	Two-Layer k-ε model with 'All y+ Wall Treatment'
Solver	Coupled Flow and Energy with 2nd order implicit method
Rotation	Relative Reference Frame
Time modeling	Unsteady
Time step	$[5E-5...5E-6]$ s

© Springer Fachmedien Wiesbaden GmbH, part of Springer Nature 2020
A. Kächele, *Turbocharger Integration into Multidimensional Engine Simulations to Enable Transient Load Cases*, Wissenschaftliche Reihe Fahrzeugtechnik Universität Stuttgart, https://doi.org/10.1007/978-3-658-28786-3

Table A1.3: Models chosen in GT-Power for the calculation of VHGTB and
the two-cylinder engine

Working Fluid	Simplified, ideal gas with the properties of nitrogen
Solver	Explicit, Runge-Kutta
Spatial discretization length	$[1...3mm]$

Table A1.4: Approaches for an integrated turbocharger

	Appr. 1	Appr. 2	Appr. 3	Appr. 4
Cylinder	3D-CFD	3D-CFD	3D-CFD	3D-CFD
Manifold after compressor	Partially 3D-CFD	3D-CFD	3D-CFD	3D-CFD
Manifold before turbine	Partially 3D-CFD	3D-CFD	3D-CFD	3D-CFD
Turbocharger	-	Map	Hybrid	3D-CFD
Manifold before compressor	-	3D-CFD	3D-CFD	3D-CFD
Manifold after turbine	-	3D-CFD	3D-CFD	3D-CFD
Boundary conditions	unsteady	static	static	static
Mass balance	-	Map	3D-CFD	3D-CFD
Energy balance	-	Map	Map	3D-CFD

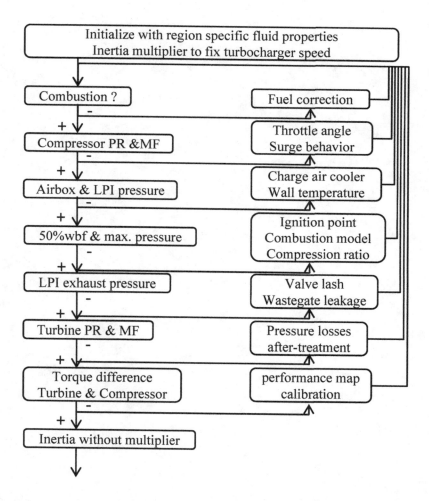

Figure A1.1: Calibration procedure for QuickSim with 0D-Turbocharger

Printed in the United States
By Bookmasters